S0-BFP-025

# Human Aspects in Office Automation

Elsevier series in office automation, 1

# Human Aspects in Office Automation

edited by

BARBARA G.F. COHEN

*Division of Biomedical and Behavioral Science, National Institute for Occupational Safety and Health, Cincinnati, Ohio, U.S.A.*

**Elsevier**
**Amsterdam – Oxford – New York – Tokyo 1984**

ELSEVIER SCIENCE PUBLISHERS B.V.
Molenwerf 1
P.O. Box 211, 1000 AE Amsterdam, The Netherlands

*Distributors for the United States and Canada:*

ELSEVIER SCIENCE PUBLISHING COMPANY INC.
52, Vanderbilt Avenue
New York, NY 10017, U.S.A.

ISBN 0-444-42327-3 (Vol. 1)
ISBN 0-444-42328-1 (Series)



Printed in The Netherlands

To clerical workers everywhere and to the people who aspire to provide them a healthy, humanized workplace.

CONTENTS

# PREFACE

## PUBLIC HEALTH AND THE WORKPLACE

WILLIAM C. WATSON, JR.
Centers for Disease Control
Atlanta, Georgia

One of our highest priorities in public health is the protection of millions of Americans during their working lives. NIOSH and the Centers for Disease Control provide the critical elements in a national strategy for public health. That strategy includes major emphasis on occupational and environmental health during the remaining decades of this century. Together, the health professionals and workers of the United States face problems and opportunities not previously imagined.

The earliest hazards in the workplace probably had to do with injuries inflicted by the wild animals hunted for food in the forests. When man had established his superiority in that environment, the challenges became more complex. It was many more years before the office work force became a factor in the equation of health and occupation. In Dickens' book, _David Copperfield_, no one measured the height of Uriah Heep's work stool out of concern for his back.

Some of the first occupational hazards to appear in the literature are still relevant. Hippocrates was concerned about lead toxicity in the mining industry in the fourth century B. C. The first recorded case of asbestosis was reported at Charing Cross Hospital in 1900. The victim was a lone survivor of ten who had worked in the carding room of an asbestos mill in 1890. As recently as 1933, a champion of worker health pointed out that one case of silicosis often costs more than the $10,000 needed for a company's prevention program.

More than 100 years ago, it became apparent to some that prevention was the answer to the economic penalties and health hazards of dangerous working conditions. In the 18th century, a noted English scientist and investigator advocated that benevolence toward working people "be directed to the prevention, rather than to the relief of the evils." The first evidence of government legislation to protect the health of working men appeared with the English Factory Acts of 1833.

In 1911, occupational disease reporting began in the U. S. with the enactment of laws in six states requiring physicians to report cases of anthrax, compressed air illness, lead poisoning, and other toxic exposures.

John W. Trask, a discerning epidemologist and Assistant Surgeon General of the U. S. in 1914, realized the inadequacies of early reporting efforts. He said:

> No epidemiologist would think of attempting to control an outbreak of yellow fever....without inaugurating a dependable system whereby he would receive prompt and accurate information of the occurrence of cases. It is just as impossible to effectively control industrial lead poisoning, or any other preventable disease without a knowledge of the occurrence of cases.

Organized prevention was still in the future. Lacking definite statistics about industrial morbidity and mortality, a group of public health specialists completed a study of manufacturing plants in a major industrial area of the U. S. in 1934. Their purpose was to determine the status of industrial hygiene in that area. Even then, they found that nearly 20% of the workers were still using a "common drinking cup" and that 13% used a "common towel." More interestingly, they found that at age 20, industrial workers had a life expectancy that was seven years shorter than their counterparts working in offices. The industrial worker at 20 could expect to live to be 62 years old. The office worker could look toward 69 years of age.

Now after nearly 50 years, we can look back with pride on the progress that has been made. In 1979 there were 13,200 job-related, accidental deaths in the United States--a 71% reduction in work deaths since 1912. That decrease has occurred in spite of a work force more than double in size and a ninefold increase in production. In addition to accidental deaths, there are an estimated 390,000 occupational diseases reported each year. This is probably the tip of an iceberg. Illnesses related to occupations are hard to pin down and may not show up until many years after exposure.

In the secretarial and clerical fields, there are many intangibles that relate to health or disability in the workplace. For many years, study and control measures of occupational disability and death were concentrated in factories and mines, for there the dangers were visible. The office was assumed to be a safe place to work. Future efforts must be devoted to making it a safer place to work.

Today clerical workers spend on the average more than 38 hours per week at their jobs. New technology, designed for efficiency, speed, and accuracy may be producing pressures that now, or later, translate into ill health or disability. For example, clerical video display terminal operators reported more visual, musculoskeletal, and emotional problems than others in a comparative group of professional VDT users and a control group who did not work with VDTs. Results of the study indicate that health complaints were traced to a variety of interacting factors, including workplace design, job structure, and the general working environment.

Duplicating machines have been investigated as sources of exposure to toxic substances; secretarial and clerical workers have been exposed to carbon monoxide in the workplace; workers in similar jobs near laboratory areas have had to be protected from harmful fumes introduced into the environment because ventilation systems were not adequate for handling toxins.

Our techniques for coping with the health needs of workers have been modernized. But the hazards of the late 20th century have become more complicated, less definable, and more pressing. They include new toxic substances, invisible carcinogens, differing levels of radiation, and increased environmental pressures. We have just begun to address the serious questions of possible health hazards to men and women working during their reproductive years. What conditions in their places of work might be potentially damaging to their unborn children? The stress of working in an environment insensitive to such issues as sexual harassment or special demands on working mothers can take an unacceptable toll on the health of many clerical and secretarial employees.

In 1933, the U. S. Bureau of Labor Statistics Bulletin listed 94 groups of industrial poisons. Today there are more than 35,000 substances listed in the Registry of Toxic Effects of Chemical Substances. One in four U. S. workers, close to 21 million, may be exposed part- or full-time to OSHA-regulated hazardous substances.

In spite of it all, today's worker labors in the best working environment in history. Industry and government have both recognized the need for a healthful work area as a vital contribution to the economy and the future of our nation. Today's worker is better educated about his health, better protected by his employer, and better provided for in the event of accident or illness, than ever before.

What are our options for the future? The most basic is to apply what is already known about the prevention of morbidity and mortality to our efforts in the workplace. We have tools to do this successfully. We need systematic surveillance systems in each area of industry to determine where there is unnecessary morbidity and mortality. Such systems are comprised of three basic steps in the surveillance cycle: (1) data collection, (2) analyses and conclusions based on that data, and (3) response to the conclusions.

There are surveillance systems for a number of infectious diseases now, but we do not have a national surveillance system for occupational hazards. To be effective in occupational health, these systems must collect data that enable action. It is self-defeating to ask industry for data that cannot be utilized.

Data collection should be systematized to the extent possible. Standardized forms, a reliable time frame, and similar systematizing make the collection of data easier, more effective, and more palatable. The more automatic the system is, the less likely it is that significant information will be overlooked. Outbreak investigations, or health hazard investigations, provide a measurement of how effective the formal reporting system is.

After data are collected, they must be analyzed. The analysis of data is a search for truth. Particularly in occupational health, we must strive for objectivity in our analyses.

Response to collected data must occur on two levels. First, information must be disseminated to those who need it in order to take action. Second, action must take place based on the analysis. Whether we are dealing with salmonellosis, Legionnaires' disease, or asbestosis, the surveillance arc is not completed until proper public health action has been taken to modify the chain of events.

The basic idea of such a system is certainly not new. But doing national disease surveillance systematically and on a continuing basis is relatively new in the U. S. Such studies started with malaria and have become an integral part of infectious disease control. This type of surveillance is just beginning in environmental and occupational health. We must expand the current system of infectious disease reporting in order to provide the occupational health field with the same high quality information that has been available on infectious diseases for the past three decades.

We need to set specific objectives in the area of occupational health. Most importantly we need to strengthen our capability to respond to the problems we identify.

We can recommend standards, improve control technology, give broader clarification of emerging information, and develop new methods for prevention of work-related health problems.

Finally, we need to more closely integrate occupational prevention with the total picture of disease prevention. We know that in many cases, the risks of the workplace are aggravated by the synergistic influence of other health risks.

Professionals must continue to improve the work environment, but they also need to involve the individual, as much as possible, in his or her own health protection. Today, the working man or woman can make lifestyle decisions that will add more years to a lifetime than any scientific discoveries.

Recent studies have shown that the chronic, degenerative diseases once considered inevitable will yield to prevention. Lester Breslow of the University of California has done studies to show that most people can make daily decisions that will influence their longevity more than all of today's medicine. By not smoking, using alcohol in moderation, eating a proper diet, and getting the right amount of exercise and sleep, a person in his 40's can expect to live 10 or 11 years longer than a person who does not make the same decisions.

Many industries are helping employees to implement these lifestyle decisions by providing access to smoking cessation clinics, to alcohol control programs, to weight reduction programs, and to stress counseling programs. One company estimates savings of approximately $419,000 a year by reducing absenteeism from alcoholism alone.

Occupational sites are good places for such programs. The Metropolitan Life Insurance Company estimated in 1979 that 90 to 95% of workers will participate in multi-phasic screening programs in industry. The same program offered free-of-charge in a community would do well to achieve 30% participation.

We are at a turning point in history. Our efforts in occupational health today must be the foundation for health protection and disease prevention in the workplace for the 21st century and beyond.

We now have a national strategy for health promotion and disease prevention that will strengthen our efforts in all areas of prevention. We have data to enable working men and women to make informed lifestyle decisions to protect themselves and we have means of communicating that information to them. We are well on the way to providing improved health and well-being for future generations of Americans.

We have much studied and much perfected, of late, the great civilized invention of the division of labour; only we give it a false name. It is not, truly speaking, the labour that is divided; but the men:
- Divided into mere segments of men - broken into small fragments and crumbs of life; so that all the little piece of intelligence that is left in a man is not enough to make a pin, or a nail, but exhausts itself in making the point of a pin or the head of a nail. Now it is a good and desirable thing, truly, to make many pins in a day; but if we could only see with what crystal sand their points were polished - sand of human soul, much to be magnified before it can be discerned for what it is - we should think there might be some loss in it also.
John Ruskin, The Stones of Venice, 1851 (p. 165, Vol. 2, 5th Ed., London, 1983)

# INTRODUCTION TO HUMAN ASPECTS OF OFFICE AUTOMATION

B. G. F. Cohen

Working can and should be beneficial to one's health and well-being. But if the human element is not considered in the workplace, the potential benefits of working are lost. Office work has traditionally been depicted as neat, clean and free from factors that threaten safety and health. Typically, the occupational health problems of clerical workers have been dismissed as insignificant or trivial, despite research indicating that many subtle and overt concerns of office workers need to be addressed (ref 1, 2, 3, 4). The advent of automation can and should further enhance human health and well-being by decreasing the number of repetitious and tedious tasks and increasing opportunities for accomplishments and personal job satisfaction. Instead, automation in the office--the largest and fastest growing sector of the workforce--is generating more tedium and job fragmentation, which imposes additional problems to this occupational group. Job strain and associated health problems develop particularly when the worker is perceived more as a part of the automated system than as an important human being. These problems can no longer be brushed aside since scientific evidence documents occupational stress as a threat to worker health in numerous offices (ref 3, 5, 6).

What are the realities of office work stress? What impact have they had on worker health and well being? How can they be effectively managed to create a positive, productive atmosphere for employees at all levels as well as employers? Such questions are considered in this book by a distinguished and diversified group of authorities on worker health and office automation.

The material focuses mainly on five areas of working life: (1) office environmental health issues, (2) work organizational factors, (3) ergonomic aspects of the workplace, (4) physiological and psychological effects of office work, and (5) strategies for alleviating worksite stress. The challenges posed by office automation and some suggested solutions for dealing with recognized problems are discussed within each of these five areas..

The physical environment in which office workers perform their jobs deserves attention. A growing number of health complaints have been reported by office workers in the last few years, including reports of skin, eye, and throat irritations, headache, nausea, and other ailments. Also a significant increase has occurred in the number of requests for health hazard evaluations in office work areas by federal and state health agencies. Problems in office environments have been varied and extremely complex, many suggesting a form of indoor air pollution. The increasing reports of them coincide with the growing number of renovated buildings with deteriorating asbestos ceilings and insulation and with the construction of airtight, energy-efficient buildings.

Environmental problems must be evaluated as early as possible to assure worker health. Uncovering worker symptoms is essential to this process, but any resolution depends on relating these symptoms to all of many varied sources--underlying psychosocial stressors as well as the suspect physical or chemical agents which may trigger the reactions. So although removal of toxic agents from the environment is a primary objective, it is also essential to allay the fears and apprehensions accompanying physiological symptoms, whether or not the concentrations pose a serious health threat. Such human aspects at the workplace must be considered for an effective solution to any worker health problem.

Considering the human component of the workplace also requires a candid look at work organizational factors since these factors determine the relationship between the entire working environment and the employee. Organizational procedures are at the very core of worker attitudes, behaviors, motivation, and productivity. In most cases, management attitudes and behavior filter down from the top and spread through supervisory channels to envelop every facet of the workplace. When the management style is autocratic, inflexible or otherwise oppressive, it often stifles cooperation and the sharing of ideas, knowledge, and workload. Then the work atmosphere can become clouded with mutual distrust, insecurities, and ambiguous messages. Energies that could be directed to organizational goals are often wasted on survival tactics that spawn even greater distrust, insecurity, and inefficiency.

Insecurities increase considerably during periods of change such as in the process of automating office work. Organizational practices can smooth the way and facilitate the change to automation by enlisting worker participation at every phase: from the inception of the idea, to the selection and implementation of equipment, and throughout daily problem-solving procedures.

Ironically, theories of job participation were initiated in the United States, but they have not been implemented here as widely as in Europe or Japan. Indeed, the United States has been accused of having the lowest level of participative management and information system design as well as traditional adversary relations between management and labor (ref. 7) However, participative approaches have been predicted to displace the traditional organization structure by the year 2000, but with features unique to American culture and organizational setting (ref. 7).

Management/employee participation can generate ideas and solutions to mitigate negative effects of work overload or underload, promote more task control and skill utilization, and enhance feelings of individual worth. As a result, organizations will likely experience less employee sick leave, less on-the-job absenteeism, and staff turnover, and more quality production (ref. 8).

Another concern for the human aspect of office automation is the interface between workers and their work stations: the ergonomics. Interest in the ergonomic design of the office has been stimulated by the escalating health complaints of video display terminal (VDT) operators. Correcting the tangible ergonomic causes of employee discomfort (poorly designed chairs, desks, illumination, etc.) has naturally been a more popular approach to employees' complaints than redesigning intangible organizational factors.

Understanding the way in which the human visual and musculoskeletal systems work provides the basis for constructing a well-designed work station. Equally fundamental is an understanding of the whole person, including the workers' emotional, social, motivational and psychological requirements. The best designed jobs maximize the person's unique contribution to systems and promote human dignity (R. Pew, personal communication, Human Factors Engineering Seminar, Ann Arbor, Michigan 1983). This text addresses ways of reconciling the demands of automated office systems with the following human factors: (1) Visual load, which may result in eye irritation, fatigue, and headache, (2) muscular load, associated with soreness in the muscles and joints of arms, hands, wrists, and fingers, (3) postural load, identified with muscle and joint soreness of the back, legs, and shoulders, and (4) emotional load which imposes anxiety, depression, and fatigue.

One simple task can burden all four systems simultaneously, when presented improperly. The universal task of typing can well illustrate this point when the hard copy to be transposed is poorly written. If the typist must decipher illegible or sloppy penmanship or pages crammed with insertions, the entire visual system is strained. While the eyes are straining, back, neck, and arm muscles become tense. Errors are then more likely, necessitating retyping.

In addition to suffering from tired fingers, wrists, and arms, workers may experience negative emotions as well. Staff members who would be too embarrased to submit sloppy, hard-to-read copy to a peer or supervisor seem to have no compunctions about giving that work to a secretary. This may convey to clerical personnel that they are so unimportant that it does not matter how poorly work is presented to them; their feelings and their time are depreciated. Such an emotional load minimizes self-worth, degrades self-esteem, and intensifies fatigue. The better we understand our intricate emotional, psychological and physiological systems, the more we see how stress and strain can spiral upward and make the individual more susceptible to a variety of discomforts and health complaints. Since the individual is an integral part of the work environment, every human aspect must be thoroughly considered before undertaking ergonomic workplace design.

Researchers have come to widely separate conclusions about clerical workers. One study notes them as the healthiest work group (ref. 9),and another shows female clericals as having a significantly greater incidence of heart disease than most other occupational groups (ref. 4). This discrepancy points up the need for more research on this group of workers. Although studies that include detailed measures of mental health are rare, the patterns of depression and dissatisfaction noted by mental health professionals over the years warrant immediate consideration and further clarification.

The causes or sources of health problems in an office setting are difficult to identify because of complex physical and psychological factors affecting each worker. Furthermore, the cumulative stressors that chip away at worker health daily in the office, are not as apparent as the hazards found in other types of industrial settings, yet evidence mounts to indicate that it is the chronic, sometimes invisible, hazards that contribute most to serious physiological disorders, sometimes even fatal outcomes.

However, health and well-being depend on maintaining a nominal stress or arousal level--one that is sufficient to stimulate interest and motivate accomplishment, but not so high as to impair performance. This book describes several stress reduction strategy programs that have been initiated at the worksite with good results for preventing or alleviating some of the more debilitating effects of job stress. Minimizing the impact of job stressors at work or at home is an ongoing process that requires continual joint effort and pays large dividends to both workers and employers. The techniques described here merit the attention of everyone concerned about the human aspects of office automation.

In the introductory quotation, John Ruskin warns that fragmenting tasks to save time and money produces goods but incurs human costs that may never be compensated. We must heed that warning and mind the real costs of office automation which are easily disregarded in a society swept up in the exhilaration of the latest technology. Certainly automation has the potential for eliminating repetitiousness and tedium and freeing the worker for more challenging, gratifying accomplishments. However, too often the clerical staff is simply treated as part of the machine. If we consider the human element in the process of automation, benefits will accrue to both labor and management. This volume examines the views of many on this subject--researchers, labor group leaders, management, academicians, and office workers. The initiative and energy of all these factions working together will be required if we are to prevent adverse health consequences and enjoy the benefits of office automation.

## REFERENCES


1 G. Gurin, J. Veroff, and S. Field, American View Their Mental Health, Basic Books, New York, 1960.

2 American Academy of Family Physicians, Life Style in Different Occupational Groups, Survey Report, Kansas City, Mo., 1979.

3 M.J. Smith, M.J. Colligan and J. Hurrell, Jr., A Review of NIOSH Psychological Stress Research-1977, Proceedings of Occupational Stress Conference, DHEW(NIOSH) Publication No. 78-156, 1978.

4 S.G. Haynes, M. Feinleib, Women, Work, and Coronary Heart Disease: Prospective Findings from the Framingham Heart Study. American Journal of Public Health. 70:2 (1980) 133-141.

5 C.L. Cooper and J. Marshall, Occupational Sources of Stress: A Review of the Literature Relating to Coronary Heart Disease and Mental Health, Journal of Occupational Psychology, 49 (1976) 11-28.

6 M.J. Smith, B.G.F. Cohen, L. Stammerjohn and A. Happ, An Investigation of Health Complaints and Job Stress in Video Display Operations, Human Factors, 23, (1981) 387-400.

7 H. Sackman, Problems and Promise of Participative Information System Design, in U. Briefs, C. Ciborra, and L. Schneider (Eds.) System Design, For, With and by the User, North Holland, Amsterdam, 1983.

8 J.R.P. French, The Social Environment and Mental Health, Journal of Social Issues, 19 (1963) 39-56.

9 L.M. Verbrugge, Women's Social Roles and Health in P. Berman and E. Ramey (Eds.) Women: A Developmental Perspective, NIH Publication No. 82-2298, National Institute of Child Health and Human Development, Bethesda, Md. (1982) 49-78.

# SECTION 1
## Office Environmental Health Issues

# OCCUPATIONAL HEALTH PROBLEMS IN OFFICES -- A MIXED BAG

JACQUELINE MESSITE, M. D.
Regional Consultant, NIOSH
Department of Health & Human Services, Region II
26 Federal Plaza
New York, New York

DEAN B. BAKER, M. D.
NIOSH
Department of Health & Human Services, Region II
26 Federal Plaza
New York, New York

Office workers today are becoming more concerned about their work environments. Office work environments are changing rapidly. New work systems, office designs, equipment, and materials have been introduced into the office.

The advance in information technology has brought into offices production elements that are somewhat similar to those in factories. Office workers may find themselves doing repetitive operations in noisy environments; the equipment or physical arrangements may cause discomfort. Operators of video display terminals (VDT's) report visual and musculoskeletal complaints and they may experience job stress in some highly-paced VDT operations.

The composition of carbonless paper, which is used increasingly, has resulted in varying clinical effects. The use of formaldehyde-containing materials in desk tops, walls, and window coverings, clothing, and insulation has added to the problem. Chemicals associated with the operation of duplicating and printing equipment have raised safety questions. The widespread use of asbestos and fibrous glass insulation in walls, suspended ceilings, and air ducts has also caused concern.

More recently, the construction of office buildings with sealed windows, coupled with the reduction of fresh air supplies, has resulted in poor indoor air quality and reduced humidity. The reduced air supply, which may be an energy conservation measure, has increased the potential for the accumulation of noxious agents in office air.

As a result of all of these changes, health problems in offices are now much more varied. They can range from limited complaints with easily identifiable causes to complex, chronic, and nonspecific symptoms, sometimes without identifiable toxic agents. In some cases, a specific event, such as a new odor or an illness in a fellow employee triggers a crisis of concern among the staff about the safety of the environment. The problem then assumes a social, as well as a toxicological, dynamic. This paper presents representative Health Hazard Evaluations of office complaints completed by NIOSH, highlighting the problems encountered and suggesting some lessons learned from them.

## SKIN AND EYE IRRITATION AT A SCHOOL

NIOSH received a request to evaluate complaints of skin and eye irritation believed to be related to the use of carbonless paper. Of approximately 100 office staff and faculty who had contact with the paper, only 6 had experienced symptoms. The 6 persons were among 20 supervisory faculty who had the heaviest contact with the paper. The symptoms, noted for several years, only occurred during the heavy marking periods when faculty members handled large quantities of the paper for 1 to 2 days.

The complaints were very variable, ranging from smelling an odor to experiencing a dry, scratchy throat, or itchiness of the skin. None of the other staff reported symptoms, but there was a general concern about potential health effects from handling the paper. At the time of the visit, there were no symptoms or physical findings.

Skin patch tests had been performed by an independent consulting laboratory for the manufacturer and were reported as being negative. NIOSH analyzed the paper and found no sensitizing agents or significant irritants. Sensitization was not thought to be a factor. Recommendations included limiting the contact with the paper by spreading the marking out over several days, avoiding touching one's face and eyes while working with the paper, and hand washing intermittently while working with the paper.

A follow-up telephone call indicated that the matter had been essentially relieved. This investigation was greatly facilitated by the small number of persons affected, their knowledge of the offending agent, and the limited tests that needed to be done to delineate the problem. However, what was needed here, even more than the recommendations, was the reassurance to all the workers that the paper was not hazardous to work with and there was no danger of delayed effects (ref.1).

In this case, the toxic agent was a material the employees handled; sometimes, however, the source of the complaints does not arise within the office.

## HEALTH COMPLAINTS AT A TRADING COMPANY

For example, NIOSH received a request concerning office workers in a trading company who had been experiencing intermittent episodes of dizziness, burning eyes, skin rashes, headaches, and nausea. The trading company is an office and warehouse facility sharing a building with a graphics display company that uses a bonding adhesive to laminate plastic to plywood.

Over the previous 2 years, OSHA had responded to 6 outbreaks of sudden illness among the office employees. On several of these occasions, environmental samples were collected. Low levels of solvents contained in the bonding adhesive were detected on the premises of the display company, but not in the offices of the trading company.

At the time of the visit, there were no complaints and no solvent odors were detected. Interviews of the office employees indicated a consistent pattern of symptoms including headache, inability to concentrate, and irritation of the eyes and mucous membranes. Most of the symptoms followed the presence of an odor described as "paint" or glue," associated with the intermittent use of the bonding adhesive by the display company.

The display company had been using the same bonding adhesive for 8 years without any complaints. However, a few months prior to the initial complaints, a new air conditioning system had been installed in the building. It was suggested that the problem may have been due to the use of a ceiling plenum air return with a negative pressure, which could have caused the solvent vapors to drift from the graphics plant to the trading company office.

Recommendations included replacing the ceiling plenum with a positive pressure return and increasing the amount of make-up air. The company followed through on most of the recommendations and no further complaints arose (ref. 2).

In this instance, the source of the complaints could not be verified environmentally because the exposure was transitory; however, the develoment of the identifiable odor and symptoms following the change in the ventilation system indicated the likely source and mode of transmission of the air contaminants. The source was external to the office, but the new ventilation system carried the agent into the office.

This investigation indicates the importance of inquiring about any recent changes in the office environment, including new materials or chemicals, new work processes, and changes in the ventilation system. It is essential to question the employees systematically to delineate any pattern in the complaints. This is also true in the next instance.

SKIN RASH IN AN OFFICE BUILDING

NIOSH received a request from the management of a 3-story, closed office building after 80 of 540 employees had developed an itchy, red papular rash on the extremities, neck, or abdomen. The rashes had occurred over approximately 4 months and each individual was usually affected for a week or two. However, some developed continuing, severe skin problems. The building had been occupied for 10 years with no major alterations having been made.

The complaints began gradually during the summer, but most employees were first affected during a 2-week period in the fall. Other agencies had investigated the building and reached no concensus as to the source of the problem. One suggested that the building was infested with book lice and recommended that it be fumigated. Extermination was done twice without substantial improvement. Another investigator felt that new fibrous glass insulation, installed that fall, was responsible.

Environmental sampling by NIOSH revealed extremely low levels of fibrous glass and paper fibers. The humidity in the office ranged from 12% to 20%. A NIOSH questionnaire indicated that the problem had begun before the new insulation had been installed. Several employees associated their symptoms with disruption of the suspended ceiling panels in the offices during repair work involving the return air plenums above the suspended ceilings. A NIOSH dermatologist examined several of the employees and concluded that most had been exposed to a primary particulate irritant, likely fibrous glass or mineral wool. However, several of the most severely affected persons were not representative of the general problem and included persons with dermatitis from nonoccupational causes.

NIOSH concluded that the probable source of the particulate irritant was a sprayed-on asbestos-mineral wool insulation used within the plenum above the suspended ceiling. Intermittent work involving the ceiling dislodged the insulation, resulting in exposure of the employees to the particulates. The consistently low humidity probably exacerbated the irritant effects of the particles. By focusing primarily on the most severe cases, the early investigators had not appreciated the consistent pattern of the dermatitis that was related to the work environment. The resolution of this problem was based on determining the pattern of complaints following the work on the ceiling (ref. 3).

## LONG-TERM HEALTH COMPLAINTS IN AN OFFICE BUILDING

The three investigations cited so far involved newly arising or specific complaints and the main approach was to identify agents or processes that were new or different. However, requests may involve long-term, continuing health complaints.

One such request was for an evaluation of a large office building in Manhattan after 920 of 1,300 staff signed a petition complaining of poor environmental conditions. The building has 30 stories; the top few floors are the site of a deluxe hotel with a swimming pool. The windows in the offices do not open and the air is centrally supplied and conditioned.

Since occupying the building in 1976, some employees have complained of problems with temperature and illumination. However, in 1980, the complaints increased and concern developed over possible health effects.

The NIOSH investigation revealed that the pool in the hotel had been painted several months earlier. The paint vapors had been inadvertently recirculated throughout the whole building for a short time. A crisis over the building environment had developed subsequent to this event. Over time, the focus of concern became the building in general and was not limited to exposure to the paint vapor.

A NIOSH questionnaire revealed widespread dissatisfaction with temperature, humidity, and noise in the offices. While many symptoms were reported, no consistent pattern or significant health problems were revealed. Environmental monitoring indicated no significant levels of toxic substances. The ventilation system was noted to be capable of maintaining a comfortable temperature and humidity, but the building management did not adequately respond to the environmental complaints and make the necessary adjustments. NIOSH recommended balancing the air system and procedures for more effective response to environmental complaints (ref. 4).

In this investigation, environmental problems had existed since the building had been occupied. While odors triggered concern among the employees, the basic problem was generally poor air quality. The frustration of the employees over the general conditions was triggered by the unfamiliar odor into a crisis about the safety of the building. Unfortunately, this dynamic of an underlying discontent or anxiety being triggered into a snowballing crisis is seen too often. The resolution of these problems requires dealing with the social dynamic, as well as conducting a thorough environmental investigation.

## RENOVATION TRIGGERS HEALTH COMPLAINTS

In another instance, NIOSH received a request from a university to evaluate a newly refurbished office building that had been evacuated 2 months after being occupied. The office space in question occupies the top floor of a loft building that was being converted to offices by the university.

Renovation had not been completed when the office was occupied in May 1980. Through May and June, wood walls were lacquered and carpeting and electrical wiring was completed. Windows on the north side of the building were removed and replaced using a caulking compound containing ethyl acrylate.

Within 2 weeks after moving into the office, 12 of 36 employees began to experience eye and throat irritation, headache, fatigue, nausea, and dizziness. Employees reported both acrid and sweet odors, and many dated the onset of their symptoms to when the windows had been replaced. Through June and July, more and more employees were affected by more diverse and nonspecific symptoms and the office was vacated.

NIOSH was asked to evaluate the building in September after the university, the local health department, and two private consulting firms had investigated the office environment and had not found a cause for the health complaints. NIOSH administered a questionnaire to the emloyees and conducted environmental sampling. Only minimal amounts of formaldehyde and volatile organics were found, indicating that the building was safe for re-occupancy. While there were several possible sources for organic solvents, ethyl acrylate vapors from the window caulking were likely responsible for the initial health complaints.

After completing the evaluation, NIOSH met with the administration and all the employees to explain the results and answer questions. Subsequent to this meeting, the employees returned to the office without further incident (ref. 5).

This episode had a social, as well as a toxicological, dynamic. The initial exposure triggered concern among the employees in the office. This concern was focused on the on-going construction, with the multiple odors and dust it produced. Although the employees discussed their problems informally, they had no structured mechanisms for communicating and dealing with the environmental and health complaints. As time went on, the employees became afraid of working in the building and demanded that it be evacuated. Despite many efforts by the university to evaluate toxic substances in the office, the employees remained convinced that the building was unsafe.

NIOSH was asked to determine whether the building was safe for re-occupancy. However, resolution of the overall problem required the determination of both the environmental and social factors leading to the evacuation of the building and the establishment of effective lines of communication between the administration and the office staff for dealing with environmental problems.

It was fortunate in this instance that the events had begun relatively recently and the evolution of the problem could be reconstructed. Sometimes the underlying conditions are not recognized and the focus is exclusively on the "triggering event." While a temporary answer may be found, it is likely that the problem will reappear in the future. After a time, a cycle of triggering events, followed by increased health complaints, can lead to anxiety and fear about the office. An example of this occurred in a private school that requested assistance from NIOSH because of a number of outbreaks of varying symptomatology over a 5-year period among the office staff in the administration building of the school.

In 1974, a sprinkler system had been installed requiring the drilling of holes through fibrous glass ceiling panels. Shortly thereafter, several employees in the business office complained of skin itchiness and eye irritation. These complaints did not abate and the business office was moved several times to different locations in the administration building. The office was repeatedly cleaned after each move, then painted, and later totally refurbished. Finally in 1978, the school rented 3 trailers and moved the business office into the trailers.

During the many moves of the business office, the complaints spread to other administrative, medical, and service offices. Over the 5 years, approximately 100 persons intermittently had symptoms of itchiness, tiredness, dizziness, eye irritation, nausea, rashes, dry skin, dry throats, vaginitis, and warts. Any change in working conditions, such as installation of new carpets, a relocation of or change in types of duplicating machines, or a sudden change in temperature or humidity, acted to trigger an increase in symptoms.

A number of medical assessments made on individual employees as well as groups of affected persons never established any pattern of pathology. Environmental sampling performed by consultants intermittently over several years indicated minimal amounts of fibrous glass as well as other common dusts, insignificant amounts of organic solvents, and normal varieties of bacteria and mold.

Analysis by NIOSH of dust obtained by vacuuming files of the buiness office showed insignificant amounts of fibrous glass and other dusts. Analysis of air samples identified minimal amounts of volatile organic and acidic compounds.

Interviews of employees indicated a wide range of minor medical conditions that employees incorrectly attributed to work at the school. For example, one employee attributed the occurrence of vaginitis to sitting on certain chairs at work. In the infirmary, the head nurse attributed her own symptoms of headache and nausea to a viral infection. However, a few nurses thought they had the same problem so it was then attributed to the so-called "contamination."

On the other hand, some symptoms could be attributed to the school environment. The original outbreak of skin and eye irritation was probably due to the fibrous glass exposure. Symptoms of afternoon dizziness and fatigue, as well as some of the eye and throat irritation, may have been due to the general lack of ventilation and mustiness noted in the offices and the dryness created by the general overheating of the rooms (ref. 6).

The overall impression by the NIOSH staff was that no serious toxic hazard existed at the school. Recommendations were made for better general housekeeping and improved air circulation and heating arrangements in the buildings.

In this situation, the initial complaints evolved into a cycle of continuing complaints and environmental interventions. While each crisis was supposedly precipitated by a new agent in the environment, the underlying dynamic was, in fact, generally poor air quality and unresolved anxiety about the safety of the building environment. Successful resolution of this type of problem becomes extremely difficult after the pattern of complaints becomes established. Effective lines of communication are needed among all the employees to dispel rumors quickly. Causes of the underlying anxiety need to be identified and studied. Finally, the investigator must gain the confidence of all the affected employees. The report of the investigation has to be accurate and plausible to the employees if it is to have a chance of breaking the vicious cycle and resolving the problem.

In summary, a number of Health Hazard Evaluations have been presented. The problems are varied and often extremely complex. In all instances, it is important to evaluate the problems as early as possible. The basic approach should combine environmental evaluation with structured interviews of the employees, designed to elucidate the development of the problems. The health complaints may be related to a specific agent or operation in the office or ventilation system, or be transmitted from an outside source through the ventilation system. Finding and eliminating the causative agent(s) should resolve the problem. However, at times the complaints represent a crisis of concern precipitated by a new odor or illness, but superimposed upon an underlying general dissatisfaction with the quality of the office environment. In such situations, resolution of the problem depends not only upon the identification of the triggering event, but also upon the recognition of the underlying stressors. To solve the problem, we must deal with both the social and the toxicological components.

REFERENCES

1 J. Messite and N.Fannick, Fieldston School, Bronx, New York, NIOSH HETA 80-69, 1980.
2 J. Messite, N. Fannick and S. Ruberto, Rainbow Trading Co., Long Island City, New York, HETA 80-66, 1980.
3 D. Baker and N. Fannick, Blue Cross of Northeastern N. Y., Inc., Slingerlands, New York, NIOSH HETA 81-058-1037, 1982.
4 D. Baker and N. Fannick, United Nations, New York, New York, NIOSH HETA 81-103-962, 1981.
5 D. Baker and N. Fannick, New York University, New York, New York, NIOSH HETA 80-240-855, 1981.
6 S. Daum, N. Fannick and J. Messite, Lincoln Hall, Lincolndale, New York, HETA 79-12, 1979.

# TACKLING TIGHT BUILDING SYNDROME: WHAT CAN WORKERS DO?

DAVID MICHAELS
Department of Social Medicine
Montefiore Hospital
111 E. 210th Street
Bronx, New York, 10467

As a result of recent scientific advances and increased worker and citizen awareness of environmental health issues, indoor air pollution is becoming recognized as a potential cause of discomfort, disease, and mortality. One important component of indoor air pollution-related health problems is *tight building* or *sealed building* syndrome, which can occur in structures with windows that do not open. This term describes outbreaks of similar health complaints in groups of workers or others who spend extended periods of time in *tight buildings* that rely on mechanical systems for ventilation.

The number of reported tight building syndrome outbreaks has risen dramatically in recent years. It has been estimated that as many as 30% of the requests for health hazard evaluations currently received by NIOSH are from office workers in sealed buildings. State and local government agencies report similar increases in reported *tight building* syndrome outbreaks. As increasing numbers of commercial and residential structures are being made energy efficient, reducing the amount of outside, fresh air that enters the building, it is likely that the incidence of *tight building* syndrome will rise (ref. 1, 2, 3, 4).

Most commonly, these outbreaks are characterized by a high incidence of similar, somewhat nonspecific complaints, such as upper respiratory irritation, headaches, dizziness, or drowsiness. This type of health complaint in office workers should not surprise public experts, since many of the nation's white collar workers are commonly exposed to a wide variety of substances that can result in these symptoms. Toxic levels of numerous contaminants have been reported in office environments as a result of malfunctioning or inadequate ventilation. These contaminants include central nervous system depressants, such as carbon monoxide and numerous solvents, and upper respiratory tract irritants, including nitrogen dioxide, formaldehyde, acrolein, and butyl methacrylate (ref. 2, 5).

Within offices, there are many potential sources of these contaminants, some of which are quite common. For example, exposure to formaldehyde, associated with health complaints in many closed commercial and residential structures, can result in upper respiratory irritation at levels far below the OSHA standard. Formaldehyde can be released from treated textiles and fabrics, including drapes and rugs, plywood and chipboard, and is a constituent, along with several other irritating chemicals, of cigarette smoke (ref. 5, 6). [Formaldehyde has been shown to cause cancer in laboratory animals; NIOSH recommends that it be handled as a potential occupational carcinogen and that appropriate controls be instituted to reduce worker exposure (ref. 7.).]

Another common source of indoor air pollution is office equipment, including photocopy machines, facsimile copiers, and spirit duplicators. Malfunctioning office machines are more likely to emit toxic substances than well-maintained ones, and hazardous exposures will be greater if machines are operated in poorly ventilated areas. Ventilation systems themselves may contribute to pollution-related health complaints. Poorly designed systems can bring contaminants into the office environment from elsewhere in the building or even from the outside. This can be a particular problem if chemical laboratories or garages are located in the same structure as offices (ref. 8, 9). White-collar workers are not immune to biological hazards either. Microbes that thrive in commercial building air cooling systems have been shown to be responsible for outbreaks of hypersensitivity pneumonitis in groups of office workers (ref. 10, 11).

Both public and private sector health officials have found tight building syndrome to be a problem that is particularly difficult to remedy. There are five constraints that hinder them in investigating and alleviating typical tight building complaints.

(1) As is the case with most occupational diseases, it is likely that the vast majority of tight building syndrome outbreaks are currently unrecognized and are likely to remain so in the near future. Individual workers may not realize that their health complaints are work-related, especially if they do not perceive a pattern of similar complaints among their co-workers. It is at least equally likely that their physicians will not be able to identify the cause of their problems, either. Most physicians are untrained in the fundamentals of occupational medicine and, therefore, are not able to ask the questions that are necessary to diagnose a work-related disease (ref. 12).

(2) The current period is one of regulatory retrenchment. Health and safety expenditures are being reduced under the Reagan administration and NIOSH, OSHA, and EPA will have fewer, rather than more professionals to do research, conduct inspections, and develop standards for the office environment. Notwithstanding these reductions, the potential magnitude of the office environment problem renders it a difficult one for even a well-staffed regulatory agency.

(3) There appears to be a significant psychosocial component in tight building complaints. Unpleasant working conditions, poor job design, and other stressors may all contribute to reports of indoor air pollution-related

symptoms (ref. 13). There are no standards regulating the boring and repetitive nature of many office jobs, machine-paced work on a video display terminal, or the demands of an insensitive boss, and the ability of governmental agencies to intervene in these situations is clearly limited.

(4) Tight building syndrome cannot be successfully approached utilizing the currently-enforced workplace exposure standards. Virtually all NIOSH investigations of widespread health complaints in office workers have found that the measured airborne levels of toxic contaminants are far less than the maximum concentrations allowable under law. This does not mean the contaminants are not the cause of the problem; the standards in question were formulated to provide limited protection against chronic disease, such as cancer, or an acute disease or symptom. Exposures to legally-permissible levels of toxic chemicals can result in illness or health complaints in exposed individuals (ref. 14).

(5) Finally, in those situations where workers do identify their problems as environmentally related, they often lack the awareness necessary to refer the problem to the proper agency. Since only a small percentage of office workers are unionized, white-collar workers are relatively powerless. As a result, they cannot utilize the most important resources blue-collar workers traditionally use in approachng safety and health problems.

Over the past several years, my colleagues and I have investigated numerous tight building syndrome complaints for unions and employers in the New York area. It is our conclusion that tight building syndrome can often be reduced or eliminated if the workers and employers involved investigate the problem themselves, without utilizing government assistance.

For example, we were asked to investigate the high prevalence of upper respiratory symptoms among teachers in a metropolitan area elementary school. The teachers' union held a series of lunch-hour meetings, to which we were invited, in order to discuss the problem. With our assistance, the teachers identified and mapped the work areas of those suffering the most severe symptoms. Together, we discovered that the reports of respiratory irritation correlated closely with presence of carpeting that was rarely cleaned. At our joint suggestion, the Board of Education removed the carpeting and installed a replacement made of different material, which is now cleaned regularly. While we did not identify the precise cause of the symptoms, respiratory complaints disappeared shortly after the change was made.

Based on our experience, and taking the above-mentioned constraints into account, we have designed a program to identify and eliminate or prevent outbreaks of tight building complaints. This program involves the active participation of employees who, as a resource, are often ignored in such discussions. The program also applies to employers, particularly because labor and management working together are more likely to tackle the office environment problem successfully than either working alone.

SURVEILLANCE OF WORKPLACE PATTERNS OF DISEASE

Workers can be taught to survey symptom and disease patterns in their workplace. Investigations that measure the prevalence of health complaints or symptoms can be conducted using a questionnaire or, as easily, through an interview survey. Interested workers can use this method to get a better idea of the extent, distribution, and character of the problem and identify potential causes.

Barefoot epidemiology, or population health studies conducted by nonprofessionals, is by no means a new concept.

Without training, workers have historically performed much of the most basic familiar occupational epidemiology. Chimney sweeps knew scrotal cancer was related to soot exposure long before Sir Percival Pott published his historic study on the disease in 1775; he notes in his treatise that workers traditionally referred to the disease as "soot wart" (ref. 15). More recently, it was a worker at the Bridesburg, Pennsylvania, Rohm and Haas plant who suggested that workers in building six, exposed to bis-chloromethyl ether (BCME), had an extraordinarily high lung cancer rate (ref. 16) and it was a petrochemical worker in Texas City who asked OSHA to investigate the cluster of brain cancers he noticed among his co-workers (ref. 17). Population studies that were initiated on the basis of these two reports uncovered significant excess cancer mortality in workers in these industries.

On occasion, untrained workers have performed far more complex epidemiologic investigations. A well-known worker-performed study was undertaken by Michael Bennett, the President of United Auto Workers Local 346, using the death records of his members employed at General Motors Fisher Body Die Casting and Electroplating plant in Flint, Michigan. Bennett reported excess lung cancer in his membership; a cohort mortality study by a team of professionally trained epidemiologists subsequently confirmed his findings (ref. 18).

To tackle indoor pollution, workers should be trained in survey methodology, particularly in questionnaire design and data analysis. One sort of survey workers might profitably employ involves internal comparisons. The prevalence rate of a complaint in one floor, department, or building can be compared with that of another; or a comparison might be made of temporal rates of the complaints in question (ref. 19).

As an illustration of this point, consider the case of workers in a San Francisco social service center who noticed a pattern of classic tight building syndrome complaints. With the help of the San Francisco Occupational Health

Clinic, the workers' union (a local of the Service Employees International Union) surveyed all the employees in the area where the complaints occurred. The investigators constructed a control group consisting of workers, represented by the same union, who performed identical tasks in an older building with windows that open. The survey asked workers to identify from a list of symptoms the ones they suffered. Included in the list were several symptoms, such as bladder problems and lower back pain, that were unlikely to be associated with toxic exposure in the office environment. This was done in order to judge whether the workers in the first group were merely chronic complainers. The exposed group registered affirmative responses for significantly more tight building complaints but reported approximately the same rate of unrelated symptoms (ref. 20).

## EVALUATION OF WORKPLACE ENVIRONMENTAL QUALITY

Workers can be trained to perform basic environmental evaluations of their workplaces. In fact, many are doing this already, using cigarette smoke, balloons, bubbles, and other easily acquired tools. For example, in many offices, workers have attached strips of paper to duct gratings to detect the velocity and direction of the air flow. Training in workplace environmental evaluation should include information about air flow rates, filtration, and humidification, standards and specifications, sources of environmental contamination, and the basic principles of ventilation systems. Equipped with this knowledge and with the help of their building engineer, workers can evaluate their environment in an effective, scientific manner.

The concept of worker-inspectors may appear radical in the U. S., but it has shown itself to be effective in western Europe. In England and several Scandinavian countries, a worker health and safety delegate is appointed or elected by workers in every worksite, including offices. (Five percent of the Finnish work force, for example, hold health and safety delegate positions). The worker-inspector is sent to a training program provided by the union, union and management cooperatively, the government, or academia. Training is provided in safety and health in general and in the specific hazards found at the delegates' workplaces. This is a model that we should begin to discuss in this country (ref. 21, 22).

## OFFICE DESIGN

Just as office workers can monitor office health patterns and the quality of air, they should participate in creating office designs that better fit their needs. One of the primary reasons for the existence of the tight building problem is that historically office workers have had virtually no input into the design of office buildings. Architects and corporate designers often see as their task the construction of the optimal workplace; they place productivity and aesthetics above comfort and health. For example, it is the rare modern office building in which workers have control over the air temperature, the amount of lighting, or the background music in their work areas.

Rather than designing office buildings to serve the needs of office workers, builders choose instead to erect in commercial districts throughout the world the modern, "international style" office tower (ref 23). We are all familiar with this image--clean, well-organized, uncluttered. The building appears to be a machine itself, rather than merely housing for machines and workers. A vital part of this image is the imposing glass wall, constructed with 12-ft. glass windows, hung on steel frames. Generally, these windows are designed not to open, in order to facilitate climate control and prevent accidents. Aesthetics are also a contributing factor--open windows destroy the sparkling, futuristic statement that the unbroken glass wall is designed to make.

Certainly, it is possible to design an aesthetically pleasing modern building without reducing worker comfort; but, typically office buildings are designed without office worker input and without consideration of office worker needs.

Working 40 hours a week at the same location, doing the same job, workers can undoubtedly make constructive and imaginative suggestions in office redesign that may increase both productivity and comfort. Experiments that have been undertaken by experimental psychologists in recent years have confirmed this. Seattle Government Service Administration employees, for example, who participated in the selection of furniture for their offices, were found to be much more satisfied with their offices than similar employees in a Los Angeles office who had no say in the design of their work areas (ref. 24).

## CONCLUSION

Rather than devoting even larger amounts of scarce resources to continued investigations of the increasing numbers of tight building syndrome reports, NIOSH and OSHA could have a significant impact on the problem by initiating a program that would assist workers and employers in the identification, investigation and alleviation of indoor air pollution problems. The immediate development of materials that will train workers in epidemiology, office environmental surveillance, and workplace design is important. Additionally, these agencies should assist unions representing clerical workers, the COSH's (Committee for Occupational Safety and Health, functioning in over 30 cities throughout the country), and organizations such as Working Women and 9 to 5 in developing regional training programs for their members.

Given the tools and training, workers can investigate and alleviate tight building problems. While more stringent standards concerning office building ventilation systems are called for, we cannot wait for them to be set. Workers can begin to address the problem now, before the promulgation of new standards.

ACKNOWLEDGEMENTS

I would like to express my thanks to LaVerne Campbell, Deborah Nagin and, especially, Marsha Love for their help in preparing this paper.

REFERENCES

1 U. S. General Accounting Office, Indoor air pollution: An emerging health problem, Washington, D. C., 1980.
2 Working Women Educational Fund, Warning: health hazards for office workers, Cleveland, 1981.
3 S. Budiansky, Indoor Air Pollution, Env. Sci. and Tech., 14 (1980) 1023-1027.
4 E. F. Ferrand and S. Moriates, Health aspects of indoor air pollution: social, legislative and economic considerations, Bull. N. Y. Acad. Med., 57 (1981) 1061-1066.
5 C. D. Hollowell and R. R. Miksch, Sources and concentrations of organic compounds in indoor environments, Bull. N. Y. Acad. Med., 57 (1981) 962-977.
6 H. E. Ayer and D. W. Yaeger, Irritants in cigarette smoke fumes, Am. J. Pub. H., 72 (1982) 1283-1284.
7 National Institute of Occupational Safety and Health, Formaldehyde: evidence of carcinogenicity, Current Intelligence Bulletin #34, Cincinnati, 1981.
8 Centers for Disease Control, Employee illness from underground gas and 0.1 Contamination, MMWR, 31 (1982) 451-2.
9 D. W. Drummond and F. J. Kilpatrick, Evaluating the health hazards of exhausts near intakes on buildings, presented at the APHA Annual Meeting, November 5, 1981, Los Angeles, California.
10 P. M. Arnow, et al., Early detection of hypersensitivity pneumonitis in office workers, Am. J. Med., 64 (1978) 236-242.
11 C. E. Reed, Allergic Agents, Bull. N. Y. Acad. Med., 57 (1981) 897-905.
12 B. Levy, The teaching of cccupational health in american medical schools, J. Med. Ed., 55 (1980) 18-22.
13 M. J. Colligan, The psychological effects of indoor air pollution, Bull. N.Y. Acad. Med., 57 (1981) 1014-1026.
14 American Conference of Governmental Industrial Hygienists, Documentation of the threshold limit values, 4th ed., Cincinnati, 1980.
15 M. Potter, Percival Pott's contribution to cancer research, J. Natl. Inst. Can. Mon. 10 (1973) 1-13.
16 W. S. Randall and S. D. Solomon, Building Six: Tragedy at Bridesburg, Boston, Massachusetts, 1977.
17 V. Alexander, et al., Brain cancer in petrochemical workers: A case series report, Am. J. Ind. Med., 1 (1980) 115-123.
18 M. Silverstein et al., Mortality among workers in a die-casting and electroplating plant, Scand. J. Work Env. and Health, 7:4 (1980) 156-165.
19 M. Silverstein, The case of the workplace killers: A manual for cancer detectives on the job, Detroit, Michigan, 1980.

20 M. J. Coye, Worker participation in occupational health epidemiology: A case study of and by social service workers, presented at the APHA Annual Meeting, October 23, 1980, Detroit, Michigan.
21 L. Boden and D. Wegman, Increasing OSHA's clout: Sixty million new inspectors, Working Papers, May-June 1978, pp. 43-49.
22 J. Hako, Reforms bring labor protection closer to the worker, Work, Health, Safety 1982, Helsinki, 1982.
23 H. Hitchcock and P. Johnson, The International Style, New York, 1932.
24 J. Mackower, Office Hazards: How your job can make you sick, Tilden Press, Washington, D. C., 1981.

# FIELD DESIGN AND SOLUTIONS TO OFFICE WORKER ILLNESS: AN INDUSTRIAL HYGIENE PERSPECTIVE

JAMES D. McGLOTHLIN
Hazard Evaluations and Technical Assistance Branch
National Institute for Occupational Safety and Health
4676 Columbia Parkway
Cincinnati, Ohio 45227

## INTRODUCTION

Since 1970, more than 80 indoor air quality Health Hazard Evaluations have been conducted by the National Institute for Occupational Safety and Health (NIOSH). The majority of these evaluations have been technical assistance requests from employers or managers requesting an environmental and medical evaluation of their office environment for contaminants and related employee symptoms.

The purpose of this presentation is to give an overview, from an industrial hygiene perspective, of the kinds of office buildings NIOSH has evaluated, the contaminants associated with office worker symptoms, the sources of contamination found and routes of exposure, and the types of recommendations NIOSH has made to improve air quality for office workers. An example of an indoor air quality investigation, detailing an industrial hygiene and medical approach toward identifying the source of contaminant(s), and recommendations for control, are also discussed.

## METHODS

Of the more than 80 indoor air quality office evaluations conducted by NIOSH in the past 10 years, 42 were reviewed for the type of building investigated, worker health symptoms, environmental tests and their results, conclusions, and recommendations.

## RESULTS AND DISCUSSION

Buildings where office workers had complaints about indoor air quality were generally of four categories: Federal, state and private office buildings, and schools. Routes of indoor air contamination were generally from three sources: (1) external sources such as exhausts from combustible engines, or from smokestack emissions from adjacently located factories that are drawn into office buildings through air intake units; (2) re-entrainment of building contaminants such as sewer gases or laboratory gases exhausted on the roof, then recaptured by roof intake air systems; and (3) from internal sources such as photocopying machines, furniture, building materials, and cigarette smoking. A fourth category denoting odors from an unknown source was also reported in many of the NIOSH evaluations.

External sources of office building contaminants

In general, NIOSH evaluations on external sources of indoor air quality contamination were hardest to correlate when comparing office worker symptoms and contaminant levels. Environmental evaluation and measurements for such exposure were first based upon employee interviews, then a "case" definition of symptoms, identification of an odor, a source of contamination, and direction of outside wind. For example, one NIOSH study found an airport office with airline reservationists potentially exposed to combusted jet fuel emissions that may have been brought into the reservation office through the ventilation system. In another study, automobile emissions from rush hour traffic were drawn into an office building because the offices were under negative pressure relative to outside air.

Re-entrainment of office building contaminants

Re-entrainment of contaminants generated from the same office building were easier to identify than external sources in terms of cause and effect. However, environmental concentrations of contaminants were usually very dilute after recirculation, and many of the employee illnesses could only be qualitatively associated with the exposures. Nevertheless, NIOSH industrial hygienists generally approached proving re-entrainment of contaminants in three ways: (1) through release of a fragrant odor such as mint oil in an exhaust hood, then standing by a window or air vent and trying to detect it by smell as the scent came back into the room; (2) by use of air sampling pumps and solid sorbant tubes. While this method is better than the sniff test in terms of qualitative and quantitative measurements of results, the results often depended upon environmental conditions and building activities on the day of sampling. For example, on a Friday no laboratory work is done, there are fewer people in the building, and fewer cars parked in the building garage. The third approach (3) is to use tracer gas such as sulfur hexafluoride. A specified amount is released at one point and a collector (that is gas chromatograph with electron capture detector) is set up at another point to detect and quantitate gas migration patterns. The advantage of the last method is that it is independent of contaminant conditions (such as carbon monoxide) on the day samples are taken, yet levels may be extrapolated from tracer gas results. A brief presentation of a NIOSH tracer gas study to determine re-entrainment of building contaminants is discussed later in the presentation.

Internal sources of office building contaminants

The third source of contaminants that can influence indoor air quality is generated within the confines of the office work space. NIOSH studies of this type have been very easy or very complex, depending upon the source of odors and nature of symptoms. Office copiers, blueprint machines, spirit duplicators, and telephone facsimile recorders, for example, have point source emissions, and the type of contaminants these machines generate are fairly easy to evaluate. Usually if workers are experiencing irritating upper respiratory effects from these office machines it is a matter of relocating them to an open, well-ventilated area away from work areas, or to install local exhaust. NIOSH has performed several such evaluations on photocopiers in offices and on spirit duplicators in schools.

Other internal sources of contaminants may be building materials, such as urea formaldehyde insulated walls, or fibrous glass duct insulation (skin irritation). Bacterial, fungal, and viral agents should also be considered when evaluating indoor air quality of offices. Infectious agents may live and multiply in areas such as heat pump water condensate pans or at outside air intake units.

Adequate illumination of the office work area is another key factor in evaluating office worker symptoms, especially in the absence of chemical contaminants where headache and eye fatigue are prominent symptoms. It is interesting to note that illumination measurements were rarely taken during NIOSH evaluations. However, when these measurements were taken, illumination levels were found to be deficient (generally less than 60 footcandles.) Illumination levels for office work recommended by the Illuminating Engineering Society range from 75 to 100 footcandles. It is also interesting to note that there are no Federal standards for illumination levels in office work environments.

Chemical irritants in cigarette smoke such as carbon monoxide and formaldehyde are reported in a number of indoor air quality evaluations by NIOSH as a source of eye and upper respiratory complaints among nonsmokers. In one NIOSH study, cigarette smoke was determined to be the main cause of employee illness (HETA 81-153-884). In these studies, recommendations ranged from office segregation of smokers and nonsmokers at their desks to a designated smoking area that could be used during employee breaks.

Poor office ventilation was found to be the most common complaint in the NIOSH evaluations conducted, and appeared from these reports to be a catalyst for the precipitation and perpetuation of office worker complaints and symptoms. In many instances of poor ventilation, NIOSH found that recirculated room air was usually sufficient, but fresh intake air was deficient (that is, less than 15% fresh air). In addition, floor to ceiling portable partitions, filing cabinets, and other large vertical objects also influenced air recirculating patterns in large rooms, which resulted in poor air circulation in office cubicles. Short circuiting of air supply was another common complaint when false ceilings contained both supply and return air ducts. In other studies, ventilation duct work was found to be disconnected in the false ceiling space. Dirty filters were also very common. Not only can dirty filters contribute to airborne dust, but they also decrease air flow and increase energy consumption. Good office ventilation cannot be emphasized enough since it is one of the key factors for eliminating many office worker complaints and symptoms.

Compounding the ventilation problem, many NIOSH evaluations contained frequent office worker complaints about temperature and humidity (too high or too low), causing discomfort and lethargy. Energy-saving policies in recent years have increased office temperatures to 80 degrees F in the summer and decreased winter temperatures to 65 degrees F. In many instances, NIOSH noted that offices were not equipped with humidity control devices; low humidity during winter may contribute to irritated and dry skin. In some instances, suspicion of fibrous glass duct insulation flaking off and blowing into the work area is sometimes confused with low office humidity, especially when itching and dry skin are symptoms.

Job stress, caused by regimented work schedules, performance and production (that is, number of people processed by claims officers in social service agencies, or phone calls handled per unit time by airline reservationists), and deadlines may also contribute to office employee illness. Several NIOSH evaluations have alluded to the stress component in the absence of environmental etiologies as a cause of employee illness. However, in very few studies of indoor air quality were formal evaluations of stress and disease adequately addressed. See Table 1 for a summary of environmental contaminants that may be found in "internal" office environments.

TABLE 1
Sources of "internal" office contaminants and office worker illness.

| Machines/materials | Chemicals/irritants | Recommendations |
|---|---|---|
| Photocopiers | Ozone<br>Toner (carbon black, binding agents)<br>Noise (nuisance) | Move to open and/or well-ventilated area |
| Spirit duplicators | Methyl alcohol (99%) | Local exhaust |
| Telephone facsimile Recorders | Butyl methacrylate | Move away from worker's desk to an open area |
| Blueprint machines | Ammonia | Local exhaust |
| Carbonless paper | Formaldehyde | Good general ventilation, personal hygiene |

| | | |
|---|---|---|
| Wall insulation | Urea formaldehyde | Boost ventilation, timing device, turn on ventilation early to clear air before workers come to work |
| Duct insulation | Fibrous glass<br>Nuisance dust<br>1 Respirable<br>2 Total<br>Asbestos | Maintenance and good housekeeping |
| Video display terminals | Radiation<br>Ultraviolet-UV<br>Visible | Annual maintenance, check for radiation leaks at back of machine. Also, check print/background contrast and screen glare |
| Cigarette smoke | Carbon monoxide<br>Formaldehyde<br>Oxides and nitrogen<br>Nitrosamines<br>Particulates | Segregation of smoking and nonsmoking areas, have smoking area for breaks |
| Temperature/humidity | General discomfort | Portable fans, humidifier in winter |
| Poor illumination | Eye strain, headache | Install more lighting, or purchase desk lamps |
| Heat pump water condensate | Infectious agents<br>Bacteria | Check for proper drainage. Potassium |
| Fresh air intake units | Fungal<br>Viral | permanganate-pills, check filters, clear bird roosts near air intake vents |

Indoor air quality office investigation--Department of Justice, Washington, D.C.

On December 10, 1979, NIOSH received requests from the Office of Personnel and Administration and the Antitrust Division of the Department of Justice (DOJ), Washington, DC, to investigate noxious odors in offices that had been associated with the occurrence of employee illnesses since January, 1978.

NIOSH conducted three site surveys: January 7-9, 1980; February 4-6, 1980; and June 3-5, 1980. On the first survey, environmental sampling was conducted for solvents, trace metals, and carbon monoxide; medical interviews and carboxyhemoglobin (breath analysis) tests were also conducted with all known affected employees. The second survey included characterization of the building ventilation system and an epidemiologic survey of 400 employees. In the third survey, environmental mapping of contaminants was conducted using sulfur hexafluoride as a tracer gas in the building's ventilation system. Also, continuous monitoring of carbon monoxide was conducted in the DOJ garage.

Our working hypothesis was that office workers were being exposed to intermittent concentrations of automobile exhaust emissions (carbon monoxide and hydrocarbons) from the Department of Justice garage. The tracer gas study was to demonstrate re-entrainment of contaminants from an exhaust fan in the garage to third-floor offices where most office worker complaints and symptoms were noted.

NIOSH findings showed that many offices in the DOJ building were under negative pressure and contaminants from the garage as well as the street could be re-entrained or pulled into office work areas.

NIOSH recommendations included balancing the ventilation system by putting offices under positive pressure, installing a make-up air system, relocating the DOJ post office, disconnecting the garage exhaust fan, and reassuring office workers that although they may still smell hydrocarbon emissions from time to time from the street, the concentrations (approximately 3 to 10 ppm) will not cause illness or permanent health effects. Fig. 1 shows an industrial hygiene and medical team approach toward solving an indoor air quality problem at the Department of Justice.

FIG. 1. Anatomy of an "office worker" health hazard evaluation.

December 10, 1979
REQUEST FOR ASSISTANCE FROM NIOSH

Complaint

Environmental: noxious odors
Medical: headache, eye irritation, dizziness, lethargy

Research Phase

1. Talk by telephone to requestor, safety officer at DOJ
2. Background information on work done at DOJ
3. Literature search

January 7, 1980
NIOSH ENVIRONMENTAL AND MEDICAL TEAM VISIT DOJ

OPENING CONFERENCE

Open discussion on problem with all interested parties

Environmental Investigation

1. Characterize physical state of building.
2. Major and minor renovation activities.
3. Inspection of offices.
4. View vehicle activity in garage.
5. Environmental measurements
   a. carbon monoxide
   b. formaldehyde
   c. airborne organics
   d. dust sample analysis
   e. humidity & temperature

Medical Investigation

1. Medical interview with all affected employees
2. Telephone interview with past affected employees.

Compare Notes
Distill Data
INTERIM REPORT #1

1. CONCLUSION:

No clear environmental or medical patterns to delineate a common source of exposure.

RECOMMENDATIONS

1. Streamline reporting system of employee illness.
2. Better documentation of employee illness.
3. Better communications to employees of renovation work being done in building.

2. FUTURE ACTION:

Environmental

1. Ventilation survey
2. Characterize air movement throughout the building
3. Measure air contaminants

Medical

1. Medical questionnaire random distribution to all floors of DOJ
2. Case-control study (Subgroup)
3. Pre-and post-shift breath tests for carboxyhemoglobin

FEBRUARY 4, 1981 NIOSH TEAM RETURNS FOR FOLLOW-UP VISIT

Environmental

1. Characterize ventilation system and air movement in building
2. Measure air for contaminants

Medical

1. Epidemiological study (questionnaire in hand-out)
2. Medical interviews (questionnaire in hand-out)
3. Breath analysis

Compare Notes
Distill Data
FINDINGS

Medical

1. 411 questionnaires handed out
   216 returned
   36 employees reported symptoms

   Symptoms most often reported were eye, nose, throat irritation, headache and fatigue
2. Symptoms attributed to poor ventilation, inadequate temperature control, automobile fumes
3. Symptoms reported on all 8 floors
4. Highest reporting of symptoms was The Justice Management Division and the Antitrust Division

Environmental

1. Air flow to rooms inadequate
2. Some offices under negative pressure
3. Ventilation system unbalanced as a whole
4. Air measurement for contaminants are negative

Case-Control Study

21 - interviewed with symptoms
34 - other employees were interviewed
who occupied nearby offices

1. Most frequent symptoms reported among cases were headache or lightheadedness, eye and throat irritation, stuffy runny nose, lethargy and fatigue.
2. No significant difference between cases and controls with regard to age, sex, race, marital status, government service rating, workload, frequency of travel, attitude about job environment or satisfaction, medical history, use of medication, or smoking habits.

Breath analysis tests for carboxyhemoglobin showed no significant difference between morning and afternoon levels and were within normal limits for cases and controls.

CONCLUSIONS

1. Work-related symptoms not limited to any floor or work division.
2. No time related cluster of cases or any consistency in duration of illness.
3. Symptoms most reported were headache, lightheadedness, mucous membrane irritation.
4. The high frequency of complaints were related to heating and air conditioning system, auto fumes, inadequate ventilation.
5. Case Control study showed no specific risk factors.
6. Ventilation system in need of improvement.

RECOMMENDATIONS

Specific ventilation improvements.

FUTURE ACTION

Tracer gas study (sulfur hexafluoride)

Release gas in garage and try to detect the gas in offices where employees had symptoms or complained of odors.

SUMMARY

In summary, an industrial hygiene approach toward evaluating indoor air quality is to:

1. Find the source and route of contamination be it:
   - External, brought into building through outside ventilation systems;
   - Re-entrainment, automobile emissions, laboratory hoods, or sewer gas vents located on the roof; check to see if exhaust vents are in close proximity to air intake units;
   - Internal, point source emission such as copying machines, duplicators, and the like; examine ceiling tile and duct insulation for irritating agents such as fibrous glass and mineral wool; note formaldehyde; check cigarette smoking patterns in offices.

2. Survey the indoor air quality of the office through the use of calorimetric indicator tubes. Examples are carbon monoxide, carbon dioxide, and ozone. Also, check for any excessive dusts and noise sources.

3. Ventilation:
   - Check air flow of ventilation system and compare it to what it should be; ventilation charts are usually available from building engineers;
   - Check make-up air, and compare to what it should be;
   - Check filters;
   - Use smoke tubes in offices to check for air movement, especially around office partitions and workers' desks.

4. Note employee symptoms, especially if there is anything unusual such as neurological disorders. Note any cluster of cases, especially with respect to time, and if an environmental agent can be associated with symptoms.

5. Get a timely response to office workers and management about findings and recommendations.

## SECTION 2
## Work Organizational Factors

# ORGANIZATIONAL FACTORS AFFECTING STRESS IN THE CLERICAL WORKER

Barbara G.F. Cohen
Division of Biomedical and Behavioral Science
National Institute for Occupational Safety and Health
Cincinnati, Ohio

The organization an individual works in can be the most important vehicle for enhancing worker self-esteem through the achievement of work satisfaction. Palmore and and Jeffers (ref. 1) found job satisfaction to be the single most relevant determinant of longevity in their longitudinal study of elderly men. Even more significant than their eating, drinking, or smoking habits, the most salient factor found for living a long life was whether or not a person liked his job. Most waking hours are spent doing or thinking about the job. In this society, our very identity is tied up in what we do and where we work. Social as well as professional introductions are comprised of name and occupation. Thus, a good job does not merely mean good pay and the means with which to live in the "right" area, go to the "right schools, etc. It represents status; this, in turn, influences how one is treated by fellow workers, supervisors, subordinants, friends, and family and, consequently, how an individual perceives himself or herself. It is clear that one's perception of status is not a trivial matter. Research studies have associated lower-status occupations with increased levels of coronary heart disease (ref. 2, 3) and poorer mental health (ref. 4). These relations were particularly evident where work was repetitive, or involved only slight amounts of physical activity.

Low status, low pay, repetitive work and low levels of physical activity on the job are the typical working conditions of many of the 15 to 18 million clerical workers and secretaries in the U.S. The boredom, lack of challenge, lack of chance to get ahead, and lack of direct involvement with organizational goals that tend to characterize such occupations can produce high rates of stress and related diseases.

Stress reduction methods that individuals can learn and practice are discussed elsewhere in this book. Although studies are described in which workers have achieved success in lessening the ill effects of daily stressors, often stress and dissatisfaction go beyond the individual: the entire organization can suffer stress and strain. Therefore, remedies must be applied to the work establishment itself rather than just to the individual employees. A stressed workplace is characterized by low morale, high absenteeism, and high turnover, compounded by difficulties in recruiting new employees. Typically, dissatisfaction and moods of apathy or hostility are chronic throughout the organization. The continual settling of grievances usurps the energy available to achieve the kind of output that enables an individual to find meaning and pride in what he or she does for a living. This chapter discusses some pertinent organizational factors that lead to either good or to ill health and dysfunction of the office worker and potentially of the entire workplace and offers suggestions to minimize organizational stress.

Fundamental to the creation of a healthy atmosphere in any organization is the daily overriding respect that is shown for each employee as a total person regardless of status or tasks performed. Such respect is communicated by the continuous open sharing of information on policies and happenings that affect the organization as a whole, in addition to any one section in particular. This sharing not only dispels rumors that tend to arise in place of factual information, it also serves to provide a sense of belonging to all personnel. Instead, the this-is-none-of-your-business attitude is replaced by the more cooperative spirit of we-need-to-do-this or we-have-accomplished-that experienced by workers who feel they actually are a part of the organization. Entrusting the staff with first-hand information concerning the work establishment begets workers' trust of management in return.

Organizational behavior is instrumental in affecting morale. It has already been shown that employees who feel that management values their opinions are likely to be more satisfied with their working conditions and exhibit positive attitudes towards innovations (ref. 5). In an era in which continual innovation and change are inevitable, it is essential that the organization concern itself with those individuals directly affected by the changes.

One change that is particularly important is automation. In their striving for job status and fulfillment, most workers hope to further their education, skills, and expectations. However, work automation (as currently conceived) tends to generate forces that act in an opposite direction, leading to routinization, simplification, and fragmentation of jobs, thereby limiting the use of education and denying subsequent fulfillment of aspirations. The large number of individuals who are prevented from using acquired education and/or skills in their jobs is going to escalate wherever computerization processes simplify tasks without considering the people performing them. Computerization of the office derives from the effort to improve the organization's efficiency. However, when the organization fails to utilize its workers' skills and abilities, it is not only as inefficient as if it neglected to use existing computer capabilities (ref. 6), it also interjects a significant source of frustration and stress into the work environment.

Therefore, the clerical worker must be involved, not as a commodity, but as an intelligent human being. Involvement is generated by promoting job participation. Job participation, as defined by Caplan and others (ref. 7), is the influence a worker has on shared decisions that involve that worker. The amount of influence exercised parallels the amount of control the employee has over the job tasks. The importance of job participation in the mental and physical health of the worker has been well demonstrated in the scientific literature (ref. 8, 9).

Cooper and Marshall (ref. 10) have recently reviewed much of this literature. They indicate that workers who perceive themselves as participants in their work are likely to be more satisfied, experience better supervisor/subordinate relationships, and have lower job turnover. Correspondingly, jobs with few opportunities for job participation foster job-related feelings of threat, depression, low self-esteem, higher absenteeism, and more intent to quit the job. Margolis and others (ref. 11) indicated the most consistent significant predictor of strain and job related stress was a total lack of job participation. Buck (ref. 12) also found that workers reporting the most pressure perceived their organization as prohibiting participation in decision-making. Cooper and Marshall's (ref. 10) review concluded that the less job participation experienced, the more physical and mental health risks result.

Excluding workers who spend 40 hours a week on their jobs from having input regarding that job not only prohibits job participation, it is also a prime example of underutilization of corporate resources and a source of worker frustration. Typically, worker underutilization consists of requiring tasks that are too easy, such as the simple, repetitive, fragmented tasks that characterize the work of many clerical positions. Research studies indicate that clerical workers whose jobs are low in technological complexity, mental effort, or contact with the public are more likely to demonstrate higher levels of physiological and psychological stress and strain (ref. 13, 14). It has been demonstrated that boring, repetitive tasks performed under postural constraints or limited mobility can lead to higher noradrenaline output and impaired health and well-being, whereas mobility is correlated with calmer, more positive feelings (ref. 15).

Often, boring repetitive tasks (such as typing the same document a dozen times) are produced by administrative factors rather than job requirements. When numerous changes are made in a manuscript because they are "easy" to accomplish on the word processing equipment, they can be a serious source of worker stress if not approached in the proper fashion. Such changes divert the control of workload from the typist and impose unreasonable time pressures. One might surmise that the wonderful ability of the new word processor units to make changes so easily would increase productivity. Unfortunately, too often what is increased is the number of changes. It is not atypical for one clerical person to type for 10 or more staff members. When each of those staff members believes it is "no trouble" to change a word or a paragraph, after receiving the letter-perfect copy, more and more changes are requested. Thus, a great deal of extra work is generated. A lot of extra time is spent, but what are the results? Instead of more reports getting out, the same report gets out more times. Multiply this 10 or more times and the clerical worker feels frustrated and yet has no outlet for justifiable anger. Having responsibility without having control over that situation, without any authority to back up that responsibility has negative impact on one's health and well-being (ref. 16).

Office workers may experience many other instances of this lack of control over their work. Meeting deadlines is a particularly good example. When a supervisor or staff member waits until the last minute to hand a secretary a document without allowing sufficient time for the typing before the submission date, she has to drop everything else to complete the report or manuscript. Because of the nervousness and tension generated by racing the deadline, mistakes are more likely to occur entailing more time and criticism. Questions like "How far are you? Are you finished yet?" worsen the situation. In a NIOSH VDT study, it was found that whether or not workers had some flexibility or control over how their deadlines were met was an influential factor determining the amount of stress they suffered (ref. 17, 18). This is supported by American and Scandanavian studies giving both psychological and physiological support for the central role of individual control as a mediator of stress (ref. 19, 20, 21, 22). It has been pointed out that those jobs that benefit from automation are designed to incorporate increased degrees of latitude within which to regulate work arrangements (ref. 15). However, the secretary whose work plans are continually thwarted by unexpected last minute rush jobs has virtually no such control. Compounding the problem are the many interruptions (for example, taking telephone calls no one wants to answer, doing extra chores no one else wants to do) that are not only time consuming, but can be degrading. Such tasks minimize the value of the secretary's time and often devalue her intelligence. Meanwhile, the rest of the work piles up and creates a work overload.

An overload in a computer aborts the program. An overload in most systems, whether a human being or an entire organization, also leads to malfunction. Work overload is probably the most researched job stressor to date. Summarizing this research, French and Caplan (ref. 23) listed several symptoms of psychological and physical strain such as job dissatisfaction, job tension, lower self-esteem, high cholesterol levels, increased heart rate, and increased smoking. How much work constitutes an overload depends on the perception of the worker. That perception is influenced by other organizational factors such as one's role at work and one's expectations of present and future rewards for the work effort.

Role ambiguity is another problem area for clerical workers. Much empirical evidence has been collected regarding role conflicts and role ambiguity. Role conflict occurs when a person is torn between conflicting responsibilities, ideas, or conflicting orders. Examples include performing jobs that are believed to be outside the job description or against one's principles, receiving opposing messages from people in authority, or balancing the conflicting demands of responsibilities of home and work. Role ambiguity exists when a person is uncertain about what to do, what others expect him or her to do, and what the overall work objectives are.

In addition to the work role ambiguity faced by males, female workers are besieged by problems and situations unique to working women. Traditionally, it is the wife and mother who works many hours after an 8- hour work day, caring for children, cooking, cleaning, running the household, and fulfilling community responsibilities. This leaves little or no time to unwind and relax, and little or no energy to undertake further education and to acquire new skills. Some organizations allow for flexible work hours to accommodate special times for both males and females to attend PTA meetings or children's activities. However, this does not reduce the total number of hours (the overload) facing working women with families.

Exacerbating the role conflict of home and work is the role blurring that is evident at work. Sex roles are so intertwined with occupational roles that it becomes unclear which duties are job related and which chores are expected because females are thought of as errand-runners, cleaner-uppers and coffee-getters. The woman herself is torn between traditional values of how she is expected to act and how she really feels.

Over 20 years ago, Gurin and others (ref. 23) found that clerical workers reported more unhappiness, more ambivalent self-image, and more economic worries than any other white-collar group. Similarly in 1979, a study by the American Academy of Family Physicians (ref. 24), which surveyed executives, physicians, farmers, garment workers, teachers, and secretaries, revealed that secretaries reported the largest percentage of muscle aches, tension, headaches, anxiety, depression, and the smallest percentage of "no complaints." The leading cause of job dissatisfaction reported by the secretaries was poor advancement opportunities.

Providing for the advancement and the career development of the clerical and secretarial workers may be the greatest challenge facing organizations. Pervading societal thinking is the bias handed down throughout the centuries that promotes and maintains the distorted view of a woman as less intelligent, less competent, and less dedicated to her work than a man. Even the purpose of a woman working outside the home is misunderstood by the men and women in our society. It is not true that the average working woman is a part-timer working to bring in little extras her family can do without. The average secretary works full time for 30 years (ref. 25 or 26). It is projected that by the turn of this century women will average 40 years in the work force (ref. 25 or 26). Two-thirds of married female clericals' husbands earned $10,000 or less in 1979 (ref. 27). Many clerical workers are the sole financial support of their families. Yet the refusal to take working women seriously stymies promotions and maintains inequitable incomes and low status. Regardless of how many years the secretary performs her job or how well, her rewards are not evident.

Recall for a moment the example of the secretary who was given a manuscript without enough time to type it. If the paper is submitted late, the author might shift the blame and say, "I wrote it, but the secretary still has it." If, through great expenditure of effort, she makes the deadline, what is her reward? Not promotion. Not a bonus or a raise. The next time her supervisor says, "Take it to Rachel, she always gets it done." Hard work is rewarded with more hard work. No matter how good a worker, how knowledgeable or loyal, the clerical worker or secretary has minimal chance for advancing very far up the organizational ladder. Training is usually limited to skills already possessed--not to expanding skills for higher-status positions. What a waste of human resource if the only way up is to get out. Not all clerical workers even have that choice, however, since their training and their need to work lock them into a dead-end situation.

So far this portrayal of typical clerical or secretarial duties does not suggest a situation conducive to experiencing the kind of job enjoyment or satisfaction that leads to longevity, as in the case of Palmore and Jeffers' (ref. 1) male subjects. Is it the "automation syndrome" (ref. 28) that inflicts a pattern of dissatisfaction and psychosomatic complaints? Can a society so advanced as to create our miraculous machine systems know so little about human systems? Or is this society not advancing but turning back the calendar to 50 years ago, when automating the industrial workplace led to diminishing the meaning and pride in one's work and to depreciating the humanness of the worker. The management that constantly monitors the worker and gives only vague, negative feedback, can be recognized as the same type of management that in the 1930's brought about the human relations work of Elton Mayo and the Hawthorne Studies (ref. 29, 30, 31, 32). Psychologists have pointed out how using the minute tracking of performance to berate, threaten, or downgrade the employee has produced more stress, not more output (ref. 17, 33, 34, 35). So why revert to this same unhealthy management in order to increase office output? Automation should and can enhance work and the life of the worker by lessening the tedium rather increasing it. Let us explore some ideas that organizations might consider to minimize the stress experienced by an office worker.

Although job enrichment is a worthy concept, realistically not all jobs can be enriched. Each worker is an important human resource and should be treated as such by enriching his or her life. Organizations might do the following: assure employees of the importance of their contributions to the organization, explain how they fit in, and why they are needed. Acknowledge their importance as persons, not only as workers, by allowing self-enrichment courses of the individual's choice along with courses that specifically benefit the job. Encourage their expression of ideas, especially pertaining to their own jobs. Valuable insights can be gained by listening to those who perform the jobs and permitting employees the latitude to decide how their output will be met and how they will ensure the quality of that work. Effective participation is an active on-going process with tactful feedback from the organization. As mentioned in the beginning of this chapter, effective participation is one way people can identify with a company. Participating in work groups, professional clubs, or unions can also provide identity or a sense of belonging when that group promotes cohesiveness and sharing of ideas. Importantly, soliciting employees' ideas before making a change, such as automating an office, facilitates the change while minimizing fear and uncertainty.

Organizational support is another vital area to consider in reducing worker strains. Some organizations have attempted to humanize their supervision by considering not only the ergonomic factors of the work station, but also the people factors and, as a result, have provided management courses, such as active listening courses. In these, managers learn to listen to feelings as well as to words and learn how to get feedback to ensure that what a person thought he heard was what the speaker actually meant. Humanizing supervision entails positive rewards and even issuing criticism, but in a positive way, as opposed to giving feedback that downgrades a person's self-esteem and generates stress.

Organizational support includes flexible hours to alleviate conflicts between family obligations and work responsibilities. Along these same lines, special services such as day care facilities could also lessen conflicts for some workers. Overtime that intrudes into personal and family life should be limited.

Another idea worth exploring is the promotion of social support rather than social isolation on the job. Tensions can be eased and misunderstandings averted by promoting interaction between peers and between supervisor and employees. Of course, conversations can be disruptive in certain job situations. For these it is best to provide special places for breaks and for exchanging information. Taking breaks and lunch periods with friends at work (rather than alone) also promotes well-being. The supervisors or managers can take turns answering telephones at lunch time and during breaks so that clerical workers can interact with each other. Such cooperation and consideration are contagious. Starting from the top of the organization, the spirit of helping those in one's department and outside the department can spread throughout the workplace, benefiting everyone as well as the organization itself. When one secretary has a work overload and several others pitch in to help, not only is the burden lifted and work completed on time, but the feeling of camaraderie makes doing the work more pleasant. Each person, then, feels good about himself or herself. Gains derived from aiding those outside one's own work sphere are significantly enhanced when positive recognition from the immediate supervisors and overall administration is received.

In addition to such group effort, other group activities are helpful in promoting social support. Since women who need to rush home to relieve a baby-sitter, grocery shop, and fix the family meal, are often prevented from participating in any after-hours activities, lunch-hour programs that provide for new ideas or information, relaxation, entertainment or exercise might be beneficial. Wellness programs that range from exercise in equipped facilities to simple group walks around the building to employee assistance programs and stress-alleviation programs are viable ideas to promote good health. Career counseling programs should also be considered.

All these ideas to minimize the harmful effects of stress on the office worker also provide enhancement of self-esteem. Self-esteem is emphasized because it is so inextricably bound up with the physical and psychological health of every human being. Also, the impact that self-esteem has on worker health is strongly influenced by organizational factors--the utilization of capabilities, job participation, worker control over task and pace, the achievement of rewards, and the opportunities for career development. When a person can identify with the aims of the organization through participation and share in open communication, personal needs can be fulfilled and work satisfaction attained. On the other hand, if a worker is frustrated by low status and pay without hope for advancement, if bored by monotonous, repetitive, fragmented tasks, the person can incur debilitating stress-related disease and the organization itself can also suffer from stress and strain.

It is the organizational system that creates many clerical and secretarial worker problems, and it is the organization that can eliminate these problems. After the work roles are explicitly defined, in practice as well as on paper, the organization must respect each worker for the ability to know the job, the intelligence to understand priorities set, and the integrity to carry out tasks with individual discretion as to when that task can be completed or when a work break is needed. This will create the positive self-esteem that all workers need. We have created wonderful machine systems in the office but, as Toffler (ref. 36) says, we must now combine office automation with office humanization.

## REFERENCES


1 E. Palmore and F. Jeffers, Prediction of Life Span, Lexington Books, Boston, 1971.
2 S. M. Sales and J. House, Job dissatisfaction as a possible risk factor in coronary heart disease, Journal of Chronic Disease, 23, (1971).
3 J. House, The Reltionships of Intrinsic and Extrinsic Work Motivations to Occupational Stress and Coronary Heart Disease Risk, Ph.D. Dissertation, Ann Arbor, University of Michigan, Dissertation Abstracts International, 33 (1972) 2514-A.
4 A. Kornhauser, Mental Health of the Industrial Worker, John Wiley, New York, 1965.
5 R. W. Revans, Managers, men and the art of listening, in S. H. Faulkes and G. S. Prince (Eds.), Psychiatry in a Changing Society, Tavistock Publications, London, 1969.
6 W. C. Hamner and D. W. Organ, Organizational Behavior: An Applied Psychology Approach, Business Publications, Inc., Dallas, 1980.
7 R. Caplan, S. Cobb, J. French, D. Van Harrison, and S. Pinneau, Job Demands and Worker Health, National Institute for Occupational Safety and Health, Cincinnati, No. 75-160, 1975.
8 L. Coch and G. R. P. French, Overcoming resistance to change, Human Relations, 11, (1948) 513-532.
9 J. R. P. French, J. Israel and D. As, An experiment in participation in a norwegian factory, Human Relations, 13:1 (1960), 3-20.
10 C. L. Cooper and J. Marshall, Occupational sources of stress: a review of the literature relating to coronary heart disease and mental health, Journal of Occupational Psychology, 49 (1976) 11-28.
11 B. L. Margolis, W. H. Kroes, and R. P. Quinn, Job stress: an unlisted occupational hazard, Journal of Occupational Medicine, 16:10 (1974) 654-661.
12 V. Buck, Working Under Pressure, Staples, London, 1972.
13 R. Kalimo and A. Leppanen, Mental Strain in Computerized and Traditional Text Preparation, in G. Salvendy and M. Smith (Eds.) Machine-Pacing and Occupational Stress, Taylor & Francis, London, 1981.

14 A. Wisner, Work at Computer Terminals, Analysis of the Work and Ergonomic Recommendations, Laboratoire du C.N.A.M., Paris, 1978.
15 G. Johansson, G. Aronsson, and B. Lindstrom, Social psychological and neuroendocrine stress reactions in highly mechanized work, Ergonomics, 21, 1978, 583-599.
16 M. J. Smith, M. Colligan, and J. J. Hurrell, A Review of NIOSH Psychological Stress Research--1977, Proceedings of Occupational Stress Conference, DHEW (NIOSH) Publication No. 78-156, 1978.
17 M. J. Smith, B. G. F. Cohen, L. Stammerjohn and A. Happ, An investigation of health complaints and job stress in video display operations, Human Factors, 23, 1981, 387-400.
18 R. Karasek, Job demands, job decision latitude and mental strain: job redesign, Administrative Science Quarterly, 24, (1981) 285-308.
19 M. Frankenhaeuser and A. Rissler, Effects of punishment on catecholamine release and efficiency of performance, Psychopharmacologia, 17 (1970) 378-390.
20 G. Johansson, G. Aronsson and B. Lindstrom, Social psychological and neuroendocrine stress reactions in highly mechanized work, Ergonomics, 21 (1978) 583-599.
21 G. Johansson, Individual Control in Monotonous Task: Effects on Performance, Effort, and Physiological Arousal, Reports from the Department of Psychology, University of Stockholm, 1981.
22 J. R. P. French and R. D. Caplan, Organizational Stress and Individual Strain, in A. J. Morrow (Ed.), The Failure of Success, New York, 1973, pp. 30-66.
23 G. Gurin, J. Veroff, and S. Feld, Americans View Their Mental Health, Basic Books, New York, 1960.
24 American Academy of Family Physicians, Life Style in Different Occupational Groups, Survey Report, Kansas City, Mo., 1979.
25 Working Women, Race Against Time: Automation of the Office, National Association of Office Workers, Cleveland, Ohio, 1980.
26 Working Women, Vanished dreams: Age discrimination and the older worker, National Association of Office Workers, Cleveland, Ohio, 1980.
27 Women's Bureau, Twenty Facts on Women Workers, U. S. Dept. of Labor, 1979.
28 P. M. Roman and H. M. Trice, Psychiatric impairment among "middle americans," survey of work organization, Social Psychiatry, 7:3 (1972) 157-166.
29 F.J. Roethlisberger, W.G. Dickson and H.A. Wright, Management and the Worker, Harvard University Press, Cambridge, Massachusetts, 1939.
30 G.A. Pennock, Industrial research at Hawthorne, Personnel Journal 8, 1930, 296-313.
31 M.L. Putnam, Improving employee relations, Personnel Journal, 8, 1930, 314-325.
32 E. Mayo, Changing methods in Industry, Personnel Journal, 8, 1930, 326-332.
33 E. L. Thorndike, Adult Learning, New York, Macmillan Co., 1928.
34 E. C. Poulton, Environment and Human Efficiency, Springfield, Illinois, Charles C. Thomas, 1970.
35 E. A. Gammersall, M. S. Meyers, Breakthrough in job training. Harvard Business Review, 44, 1966, 62-72.
36 A. Toffler, Office of the future, The New York Times Special Supplment, 1981.

# THE IMPACT OF ORGANIZATIONAL FACTORS ON VISUAL STRAIN IN CLERICAL VDT WORK

EWA GUNNARSSON
450 Gatam 159
11632 Stockholm
Sweden

## VISUAL STRAIN IN CLERICAL VDT WORK

The rapid progress of computer technology has led to great changes in the working conditions of the clerical sector, where electronic data processing (EDP)-related techniques provide the foundation of most current rationalization.

The changeover to computer technology has had both positive and negative consequences for the affected personnel. The high frequency of visual strain reported is one work environmental problem that has become increasingly apparent in connection with the introduction of VDT work.

Since the mid-1970's, the Work Environment Fund in Sweden has financed research projects at the National Board of Occupational Safety and Health concerning work environmental problems in visually demanding jobs entailed by new technology such as VDT, microfilm, and microscopy work.

The frequency and type of symptoms of visual strain in an intensive VDT work routine were studied in a survey carried out at the Scandinavian Airlines System (SAS) booking office in Stockholm (ref. 1). Of the operators who worked mainly on screens (7.5 hours/day) 75% experienced various types of visual strain either daily or a couple of times weekly.

Unsuitable lighting and illumination conditions, the technical equipment, and the operator's vision are factors often referred to as causes of visual strain in connection with VDT work. In this study, the organization of the work and the duration of work spell were found to be crucial factors governing the occurrence of visual strain. One major finding in this investigation was that a whole day's work spell with intensive continuous VDT work is too long and causes the majority of operators various forms of visual strain.

That excessive work spells at VDT work should be avoided is further reflected in the section 6 of Directions No. 136--Reading Visual Display Terminals, published by the Swedish National Board of Occupational Safety and Health, 1979, which states:

> If eye fatigue or visual discomfort tends to develop, the work must be organized in such a way that the employee can intermittently be given periods of rest or work involving more conventional visual requirements.

During the past 2 years, several employee organizations have begun to draft requirements limiting the work spell spent at VDT's as a result not only of visual strain but also other effects, such as stress, monotony, and imposed control and rigid work procedures. The Swedish Federation of Civil Servants, for example, has demanded that the work spell at intensive VDT work should be limited to a maximum of 2 hours per person and day.

## VDT WORK AND VISUAL STRAIN IN THE SWEDISH TELECOMMUNICATIONS ADMINISTRATION

In order to obtain further basis for recommendations for work organization and work spell duration for different types of VDT work, a field experiment was carried out at the Swedish Telecommunications Administration (ref. 2). The experiment comprises part of a larger research project entitled: Work Organization and Workplace Design for Microscopic, VDT, and Microfilm Work (ASF 78/278) financed by the Swedish Work Environment Fund.

The Telecommunications Administration is planning to introduce EDP-related work routines on a steadily increasing scale. With regard to the potential consequences of increased visual strain, we were asked to study the effects of increasing the VDT workload before further computerization was undertaken and while manual work routines still were practiced.

### The sales office and the directory inquiries department

The experiment was carried out in the sales office and the directory inquiries department. The main task of the operators in the sales office is to serve customers by telephone, handling new subscriptions, removals, etc. These tasks are carried out mainly at a VDT (see Fig. 1). The work consists of periods of intensive screen work interspersed with other manual tasks.

Fig. 1. The sales office

The directory inquiries department (see Fig. 2) deals with telephone inquiries regarding changed numbers. The requested number is entered in the computer, which selects the appropriate answer from five different alternatives and displays it on the screen. The work spell is limited to 3.5 hours a day because of the monotonous nature of the task. The operator spends the rest of the day in another department where no VDT work or strenuous visual tasks are involved.

Fig. 2. The directory inquiries department.

Fifteen operators from the sales office and 30 operators from the directory inquiries department participated in the experiment. All were full-time female employees. The operators were divided into three subgroups; one younger group in the sales office (operators under 35 years) and one younger group and one older group (operators over 40 years) in the directory inquiries department.

## METHODS

Visual strain was assessed by the following methods: (1) a questionnaire about visual strain during the two experimental procedures, (2) measurements of the near-points of accommodation and convergence during the two experimental procedures (see Figs. 3 and 4), and (3) interviews, including questions about the work organization and visual strain in the normal work situation.

### Measurements of near-points of accommodation and convergence

The near-point of accommodation was measured by establishing the smallest line of letters that could be read clearly at a distance of about 30 cm along a near-point rule (see Fig. 3). The line of letters was then moved toward the base of the nose and the near-point of accommodation determined as the point at which the letters started to become blurred. The distance from this point to the base of the nose was then measured.

The near-point of convergence was measured by moving a small fixation object (in this case, the point of a pen) along a scale towards the base of the nose from a distance of 30 cm (see Fig. 4). The near-point of convergence was determined as the point at which the individual saw a double image of the object or at which one of the eyes diverged. The distance from this point to the base of the nose was then measured.

Fig. 3. Measuring the near-point of accommodation.

Fig. 4. Measuring the near-point of convergence.

Experimental procedures

Two different experimental procedures were set up for both departments (see Fig. 5). One of them was similar to the normal working procedure while the other one included more VDT work.

In the sales office the "normal" experimental procedure had a daily duration of VDT work of about 3.5 hours. The intensified experimental procedure involved a higher density of VDT work, approximately 6 hours. More telephone calls were directed to the operators so that the normally interspersed periods of manual tasks were omitted.

The "normal" experimental procedure in the directory inquiries department involved about 4 hours of VDT work while the intensified experimental procedure involved a complete day of screen work, 6 hours.

In the directory inquiries department, intensification of the VDT work consisted of prolonging the time rather than increasing the density of VDT work as in the sales office.

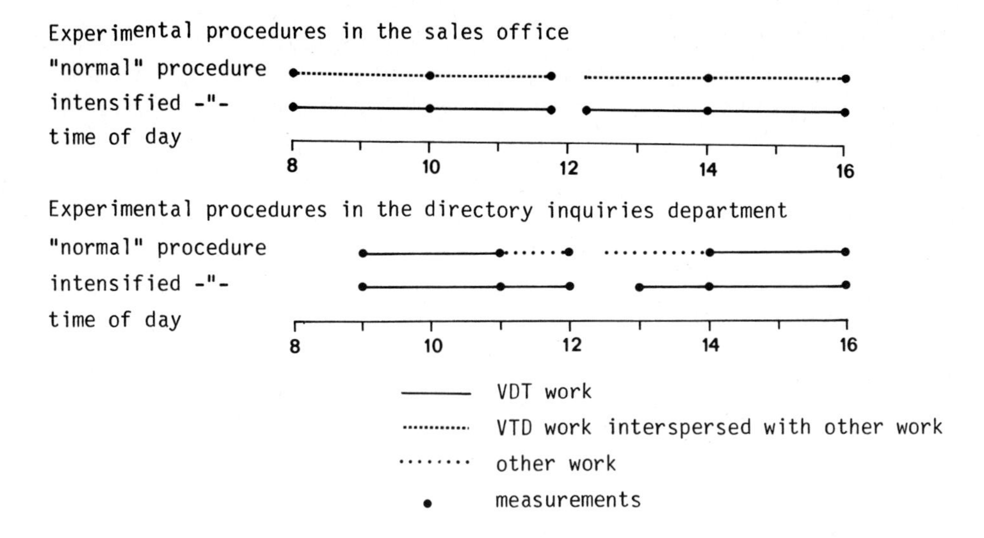


Fig. 5. Experimental procedures in the sales office and the directory inquiries department.

RESULTS

When the VDT work load is increased as in the intensified experimental procedures, the symptoms of visual strain increased in all three groups in the sales office and the directory inquiries department (see Fig. 6). The figure indicates the number of people reporting pronounced visual strain experienced during the course of the experiments. The main symptoms of visual strain reported were visual fatigue, sore eyes, and eye/headaches.

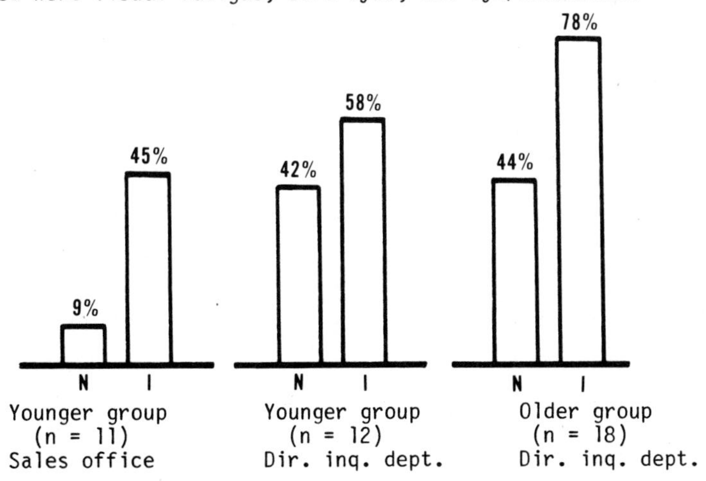


N = "normal" working procedure, I = intensified working procedure.

Fig. 6 Percentage distribution of people reporting pronounced visual strain during the course of the experiments.

Changes measured in the near-points of accommodation and convergence

The younger groups in both departments show in a similar way that the changes of the near-point of accommodation become greater as the VDT workload is increased (see Figs. 7 and 8). Comparison of before and after values for the older group in the directory inquiries department revealed that the near-point of accommodation does not change (see Fig. 9). This may be due to the fact that the stiffness of the lens, which occurs in older people after a night's rest, requires a "softening" period, which however, appears to be suppressed by VDT work.

The near-point of convergence is subject to greater and more consistent changes in all three groups (see Figs. 10, 11, and 12) during the intensified working procedure compared to the normal working procedure. The clearest difference between the two procedures appears in the younger group in the sales office (see Fig. 10), but the two other groups (see Figs. 11 and 12) showed very similar patterns.

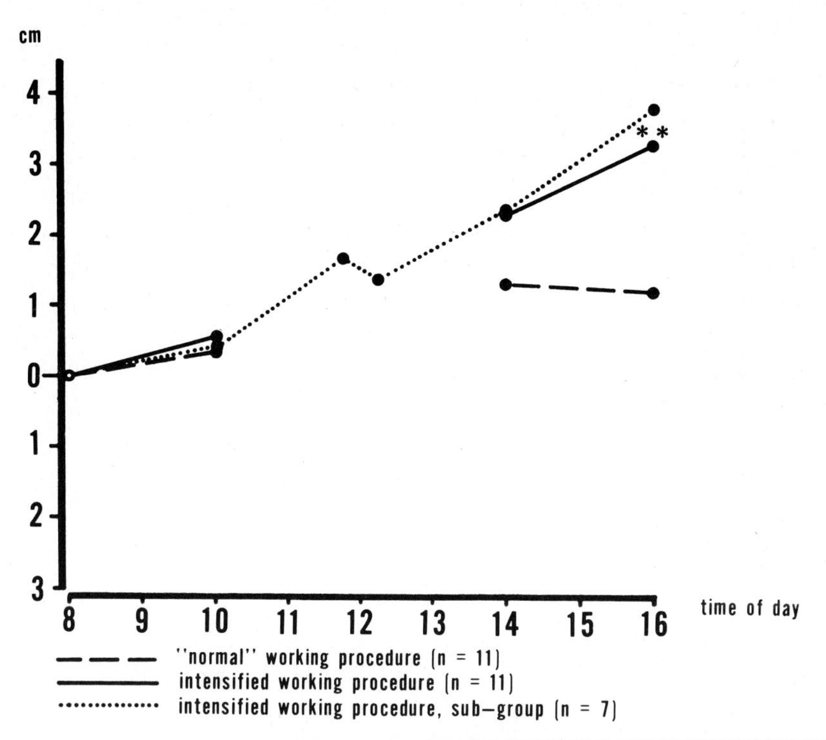


| | Normal work routine | | Intensified work routine | | Difference | | Intensified work routine subgroup |
|---|---|---|---|---|---|---|---|
| | $\bar{x}$ | s | $\bar{x}$ | s | $\bar{x}$ | s | $\bar{x}$ |
| Change | | | | | | | |
| 8 - 10 | 0.36 | 0.81 | 0.55 | 1.29 | | | 0.42 |
| 10 - before lunch | -- | -- | -- | -- | | | 1.29 |
| before-after lunch | -- | -- | -- | -- | | | -0.29 |
| after lunch - 14 | -- | -- | -- | -- | | | 1.00 |
| 14 - 16 | -0.09 | 1.70 | 1.00 | 2.45 | | | 1.43 |
| the whole day (8 - 16) | 1.27 | 3.71 | 3.36** | 3.23 | -2.09 | 5.22 | 3.85 |
| Measured values | | | | | | | |
| 8 | 13.73 | 5.14 | 12.73 | 3.82 | | | 13.29 |
| 10 | 14.09 | 5.07 | 13.27 | 3.32 | | | 13.71 |
| before lunch | -- | -- | -- | -- | | | 15.00 |
| after lunch | -- | -- | -- | -- | | | 14.71 |
| 14 | 15.09 | 5.17 | 15.09 | 5.17 | | | 15.71 |
| 16 | 15.00 | 4.63 | 16.09 | 5.28 | | | 17.14 |

** = p _0.01

Fig. 7 Changes and measured values in near-point of accommodation during the different experimental procedures for the younger group in the sales office. Measured values in cm.

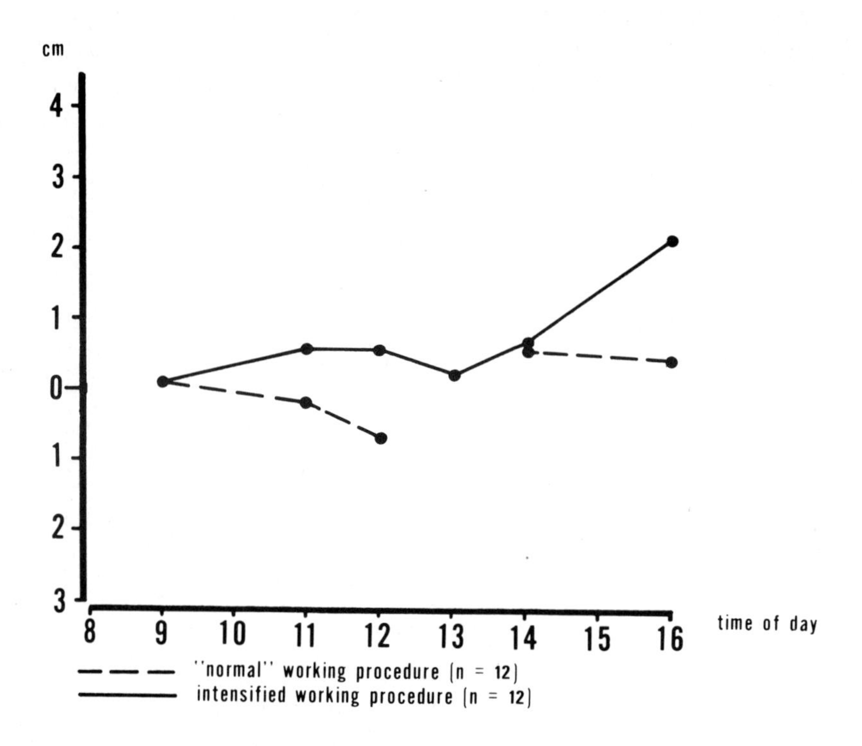


| | Normal work routine | | Intensified work routine | | Difference | |
|---|---|---|---|---|---|---|
| | $\bar{x}$ | s | $\bar{x}$ | s | $\bar{x}$ | s |
| Change | | | | | | |
| 9 - 11 | -0.25 | 1.42 | 0.50 | 1.17 | | |
| 11 - before lunch | -0.50 | 1.17 | 0.00 | 1.71 | | |
| before-after lunch | -- | -- | -0.33 | 1.67 | | |
| after lunch - 14 | -- | -- | 0.50 | 1.31 | | |
| 14 - 16 | -0.08 | 1.73 | 1.42 | 2.11 | | |
| the whole day (9 - 16) | 0.42 | 1.73 | 2.08 | 3.50 | -1.67 | |
| Measured values | | | | | | |
| 9 | 9.83 | 2.29 | 9.67 | 1.97 | | |
| 11 | 9.58 | 1.68 | 10.17 | 2.37 | | |
| before lunch | 9.08 | 2.02 | 10.17 | 2.52 | | |
| after lunch | -- | -- | 9.83 | 1.99 | | |
| 14 | 10.33 | 2.71 | 10.33 | 2.19 | | |
| 16 | 10.25 | 3.25 | 11.75 | 3.67 | | |

Fig. 8 Changes and measured values in near-point of accommodation during the different experimental procedures for the younger group in the directory inquiries department. Measured values in cm.

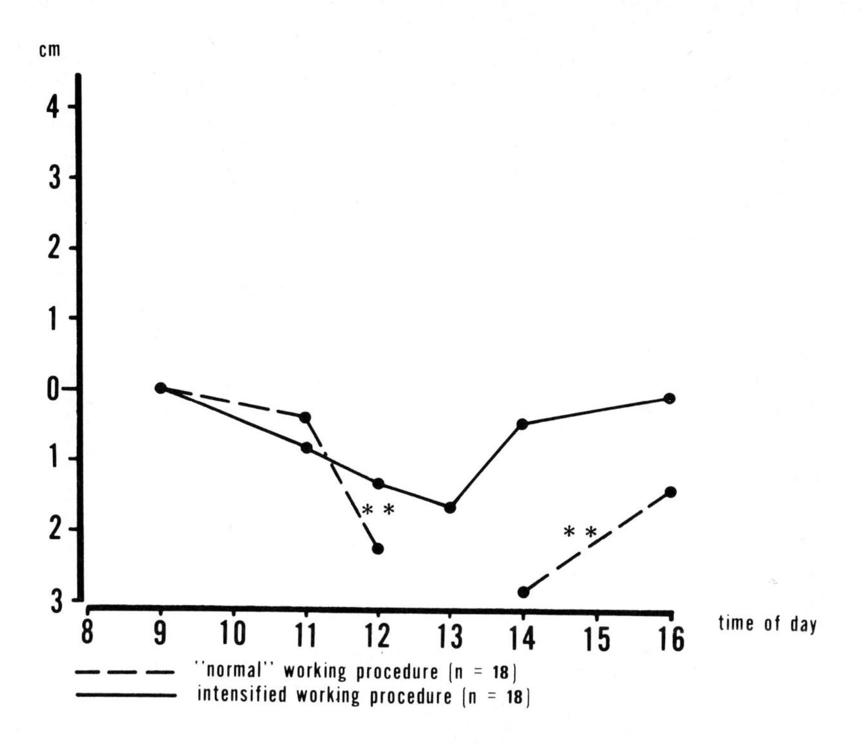


| | Normal work routine | | Intensified work routine | | Difference | |
|---|---|---|---|---|---|---|
| | $\bar{x}$ | s | $\bar{x}$ | s | $\bar{x}$ | s |
| Change | | | | | | |
| 9 - 11 | -0.39 | 2.89 | -0.83 | 2.46 | | |
| 11 - before lunch | -1.89** | 2.52 | -0.50 | 1.62 | | |
| before-after lunch | -- | -- | -0.39 | 2.66 | | |
| after lunch - 14 | -- | -- | 1.22 | 3.12 | | |
| 14 - 16 | -1.44** | 1.62 | 0.61 | 2.06 | | |
| the whole day (9 - 16) | -1.44 | 3.28 | 0.11 | 3.12 | -1.56 | |
| Measured values | | | | | | |
| 9 | 29.44 | 4.95 | 28.22 | 5.12 | | |
| 11 | 29.06 | 5.68 | 27.39 | 5.90 | | |
| before lunch | 27.16 | 5.51 | 26.89 | 5.88 | | |
| after lunch | -- | -- | 26.50 | 5.10 | | |
| 14 | 26.56 | 5.67 | 27.72 | 7.15 | | |
| 16 | 28.00 | 5.74 | 28.33 | 6.39 | | |

** = p _0.01

Fig. 9 Changes and measured values in near-point of accommodation during the different experimental procedures for the older group in the directory inquiries department. Measured values in cm.

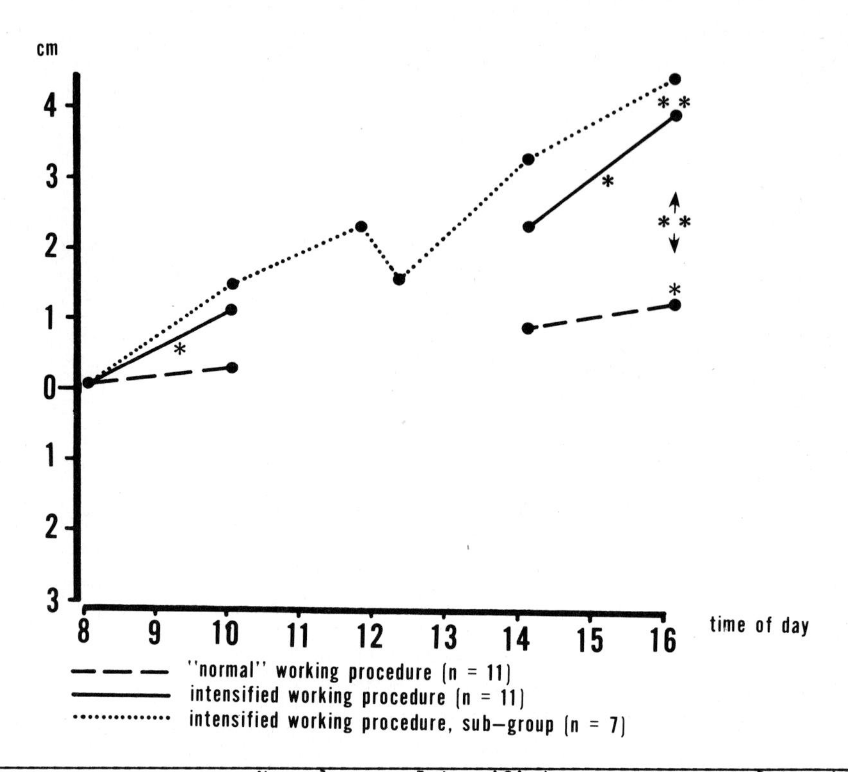


| | Normal work routine | | Intensified work routine | | Difference | | Intensified work routine subgroup |
|---|---|---|---|---|---|---|---|
| | $\bar{x}$ | s | $\bar{x}$ | s | $\bar{x}$ | s | $\bar{x}$ |
| Change | | | | | | | |
| 8 - 10 | 0.27 | 1.74 | 1.09* | 1.58 | | | 1.43 |
| 10 - before lunch | -- | -- | -- | -- | | | 0.85 |
| before-after lunch | -- | -- | -- | -- | | | -0.71 |
| after lunch - 14 | -- | -- | -- | -- | | | 1.71 |
| 14 - 16 | -0.36 | 1.21 | 1.55* | 1.69 | | | 1.15 |
| the whole day (8 - 16) | 1.27* | 1.85 | 3.91** | 2.88 | -2.64** | 3.48 | 4.43 |
| Measured values | | | | | | | |
| 8 | 6.82 | 4.71 | 6.00 | 2.79 | | | 5.43 |
| 10 | 7.03 | 4.27 | 7.09 | 3.33 | | | 6.86 |
| before lunch | -- | -- | -- | -- | | | 7.71 |
| after lunch | -- | -- | -- | -- | | | 7.00 |
| 14 | 7.73 | 4.05 | 8.36 | 4.95 | | | 8.71 |
| 16 | 8.09 | 4.09 | 9.91 | 5.11 | | | 9.86 |

* = $p \leq 0.05$
** = $p \leq 0.01$

Fig. 10 Changes and measured values in near-point of convergence during the different experimental procedures for the younger group in the sales office. Measured values in cm.

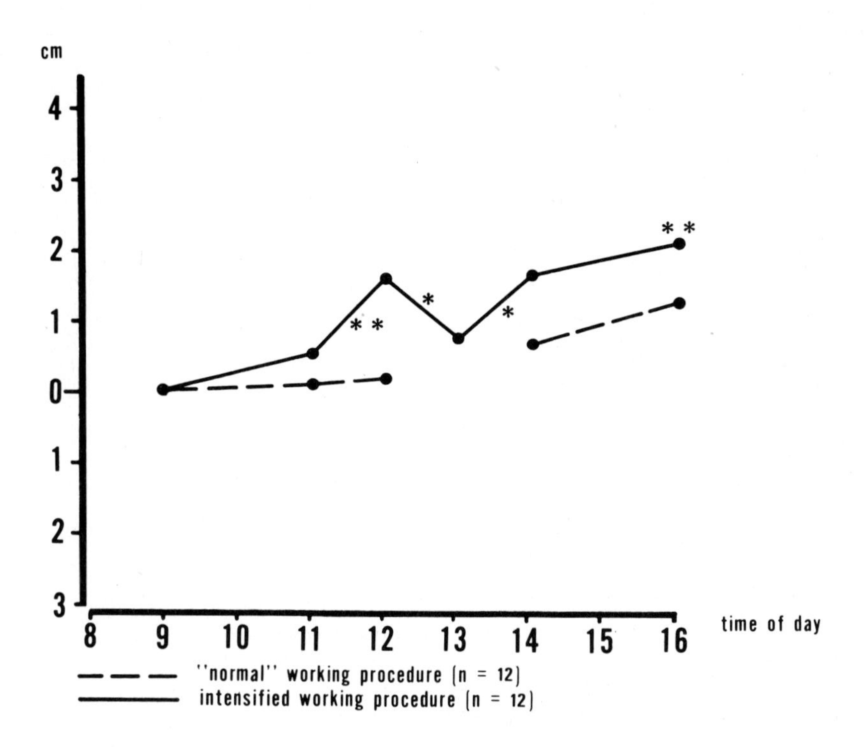


| | Normal work routine | | Intensified work routine | | Difference | |
|---|---|---|---|---|---|---|
| | $\bar{x}$ | s | $\bar{x}$ | s | $\bar{x}$ | s |
| <u>Change</u> | | | | | | |
| 9 - 11 | 0.08 | 1.00 | 0.50 | 1.24 | | |
| 11 - before lunch | 0.08 | 1.24 | 1.08** | 1.16 | | |
| before-after lunch | -- | -- | -0.83* | 1.19 | | |
| after lunch - 14 | -- | -- | 0.92* | 1.38 | | |
| 14 - 16 | 0.58 | 1.38 | 0.42 | 1.38 | | |
| the whole day (9 - 16) | 1.25 | 2.22 | 2.08** | 2.07 | -0.83 | 2.57 |
| <u>Measured values</u> | | | | | | |
| 9 | 6.75 | 2.99 | 6.42 | 2.50 | | |
| 11 | 6.83 | 2.44 | 6.92 | 2.19 | | |
| before lunch | 6.92 | 3.15 | 8.00 | 2.66 | | |
| after lunch | -- | -- | 7.17 | 2.33 | | |
| 14 | 7.42 | 2.87 | 8.08 | 3.29 | | |
| 16 | 8.00 | 2.89 | 8.50 | 3.00 | | |

* = $p \leq 0.05$
** = $p \leq 0.01$

Fig. 11 Changes and measured values in near-point of convergence during the different experimental procedures for the younger group in the directory inquiries department. Measured values in cm.

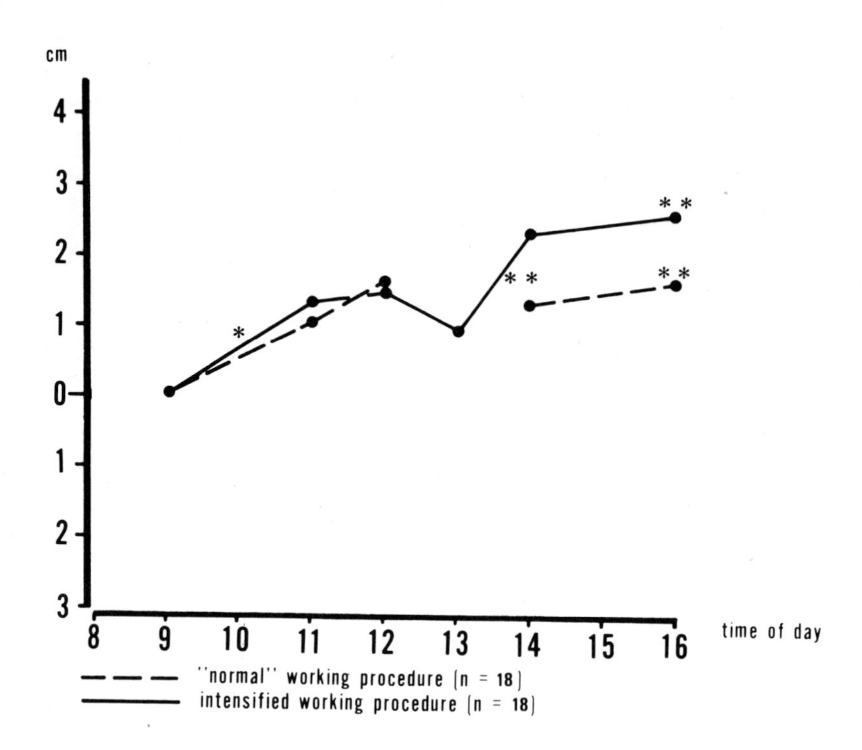


| | Normal work routine | | Intensified work routine | | Difference | |
|---|---|---|---|---|---|---|
| | $\bar{x}$ | s | $\bar{x}$ | s | $\bar{x}$ | s |
| Change | | | | | | |
| 9 - 11 | 1.06 | 2.48 | 1.33* | 2.47 | | |
| 11 - before lunch | 0.61 | 2.87 | 0.17 | 2.28 | | |
| before-after lunch | -- | -- | -0.56 | 2.97 | | |
| after lunch - 14 | -- | -- | 1.44** | 2.06 | | |
| 14 - 16 | 0.33 | 2.25 | 0.28 | 2.37 | | |
| the whole day (9 - 16) | 1.67** | 2.30 | 2.67** | 3.46 | -1.00 | 3.7 |
| Measured values | | | | | | |
| 9 | 8.00 | 3.48 | 8.11 | 3.31 | | |
| 11 | 9.06 | 4.36 | 9.44 | 4.42 | | |
| before lunch | 9.67 | 4.85 | 9.61 | 4.09 | | |
| after lunch | -- | -- | 9.06 | 4.32 | | |
| 14 | 9.33 | 4.46 | 10.50 | 4.64 | | |
| 16 | 9.67 | 4.45 | 10.78 | 5.42 | | |

* = p 0.05
** = p 0.01

Fig. 12 Changes and measured values in near-point of convergence during the different experimental procedures for the older group in the directory inquiries department. Measured values in cm.

Other investigations (ref. 3) have shown that factors such as the degree of stress, the possibility of taking a break whenever necessary, and the structural rigidity of the work situation all play an important role in the degree of visual strain experienced in VDT work. In a study on visual display units (ref. 4), 257 terminal operators were studied in various applications of the VDT. One of the major conclusions of this study was that visual strain was more frequently observed among operators whose work was highly structured, inflexible, and performed under conditions of stress. Those whose work had scope for formal breaks and varying tasks displayed fewer symptoms of visual strain. Of two groups both taking formal breaks, one also took informal breaks and reported a lower degree of visual strain than the first group.

The frequencies of visual strain reported in the interviews concerning the operators' normal working conditions are for both departments well in accordance with results obtained in earlier studies of intensive VDT work routines (see Fig. 13). The diagram clearly shows that the work spell duration is a factor governing the occurrence of visual strain.

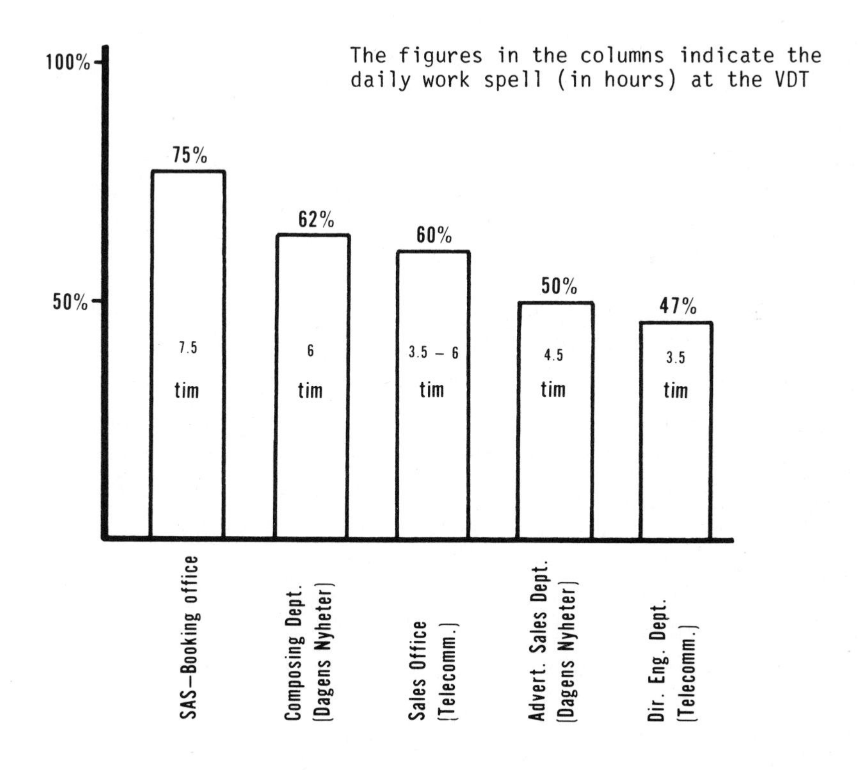


Fig. 13. Percentage distribution of visual strain in intensive VDT work routines. In addition to the results obtained from the Swedish Telecommunications Administration in the present report, SAS-Flygcity (ref. 1) and Dagens Nyheter (ref. 5).

The operators' experiences of computerization

In the sales office. most of the personnel were positive to the introduction of the computer as an aid in their working routines. Advantages quoted were the faster processing of items and improved service to the customers. The negative effect of computerization is mainly seen in that the working procedures have become increasingly controlled by the computer. The information on the screen and the order in which the images appear direct the conversations with the customer. Computer system breaks are regarded by most of the employees as very irritating and seen as a very negative effect as well.

In the directory inquiries department, the main advantage of the computerization is, quoted by all personnel, its speed in producing correct information. Furthermore, the work has become physically easier now that the handling of files and directories has been eliminated. Visual strain, monotony, and system breaks are quoted as the main disadvantages. The monotony of the work, which had already resulted in limiting the work spell duration when using the manual procedures, has hardly been alleviated by computerization, and the integration of other work tasks would be welcomed by most of the operators in the directory inquiries department.

## DISCUSSION AND CONCLUSION

One of the main purposes of the study was to obtain a basis for recommendations for suitable work spells and work organization at VDT work. Therefore, it is of particular importance to discuss the implications of the increased VDT workload in the two work routines.

When the VDT work routines are intensified, this is above all manifested in an increase in symptoms of visual strain both in the sales office and the directory inquiries department. The symptoms of visual strain increase far more in the sales office than in the directory inquiries department when the VDT workload is increased. The explanation is most probably that a combination of other factors in the work situation are interacting with the visual load and affecting the frequency of the symptoms of visual strain.

The intensified experimental procedure in the sales office implies not only an increase in the amount of time spent at the screen but also a more rigid and tightly structured work situation than in the normal experimental procedure, while the normal experimental procedure in the directory inquiries department already is rigidly structured.

The frequency of symptoms of visual strain reported in the normal work procedure in the directory inquiries department is comparable to the frequency reported in the intensified work procedure in the sales office.

Looking at the effect of an increased VDT workload on the near-point of convergence, there is a similar pattern of impairment for all three groups. The clearest difference between the two experimental procedures appears in the younger group in the sales office. For the near-point of accommodation, the changes are similar for the two young groups but differ strongly as mentioned earlier for the older group.

A relationship between the changes in the accommodation near-point and the convergence near-point is not shown under the normal experimental procedures. A clear relationship is, however, shown in the intensified experimental procedures. The possibilities to study the relationships between the near-point of accommodation, the near-point of convergence, and symptoms of visual strain were limited owing to the fact that the number of persons within a statistical group can be too small, though changes in the near-point of convergence seem to have a more pronounced relationship with the incidence of visual strain than changes of the near-point of accommodation. For all groups indicating symptoms of visual strain in the two experimental procedures, the changes of the near-point of convergence was greater than for the symptom-free groups.

This together with the homogeneous pattern of the changes in the near-point of convergence compared with the changes of the near-point of accommodation (the difference shown between the younger and the older groups) make measures of the changes in the near-point of convergence more suitable for study in further experiments as one measure of visual strain in relation to VDT work.

To conclude, organizational factors such as work spell duration and the rigidity of the work routine seem to play an important role in relation to the frequency of visual strain in VDT work. A prerequisite to alleviate negative effects such as high degrees of visual strain and sensations of stress and imposed control in the working situation is a work organization whereby the VDT is used as a well-integrated task implemented with other tasks.

## THE NEED FOR A SYSTEMATIC APPROACH TO HEALTH PROBLEMS RELATED TO VDT WORK

In general terms, the need for a more systematic approach towards health and safety problems in the working environment is expressed in the general recommendations published by the Swedish National Board of Occupational Safety and Health (ref. 6). Of particular interest for EDP related work routines are the sections that state:

> Work should be designed so as to provide the employee with an opportunity of influencing and varying the pace of work and the working methods used, and of surveying and verifying the results of his or her labors.

A shift from rigid and monotonous working procedures towards greater independence and greater vocational responsibility is essential as means of achieving greater involvement in measures of occupational safety and health.

A specific EDP-related direction integrating ergonomic as well as organizational aspects has not yet been formulated in Sweden. With the aim to prepare this type of direction, a project has been set up at the Swedish National Board of Occupational Safety and Health.

A Norwegian proposal for VDT directions states three main areas to be solved for the operators (ref. 7):

- The equipment must be adjusted to the individual operator.
- The VDT screen must be of such quality that no problems are created for the operators.
- The work must be designed in such a way that heavy physical and psychological work loads are avoided.

For data-entry work the proposal furthermore specifies:

If the VDT work consists of data-entry tasks or other routine VDT tasks the work spell duration should not exceed 50% of the total work spell. Furthermore, the VDT work spell should not exceed 2 hours without a short rest pause.

The tasks performed during the rest of the working day should be as much as possible related to the tasks at the VDT.

The background to the specifications is found in the guidelines to the proposal that emphasize the drastic increase in the strain reported after 1.5 to 2 hours in this type of work. According to the guidelines, replacing a whole day's VDT work by two part-time operators is not seen as an alternative. It is also specifically mentioned that alternating VDT work of this type with typewriting tasks is not relieving the workload to a greater extent.

The need for a more system-oriented approach to EDP work routines is also stressed by several Swedish employee organizations. The Swedish Union of Civil Servants (1982) and the Swedish Union of Insurance Employees (Working Environment in Sweden, 1981) have for example formulated extensive and detailed programs of action concerning demands related to VDT work routines.

REFERENCES


1 E. Gunnarsson and O. Ostberg, Physical and psychological work environment in a terminal based computer system. The Swedish National Board of Occupational Safety and Health, AMMF 35 (1977).

2 E. Gunnarsson and I. Soderberg, Eye strain resulting from VDT work at the Swedish Telecommunications Administration, Applied Ergonomics, 61-69, 14, 1983.

3 G. Johansson and G. Aronsson, Stress reactions in computerized administrative work, Department of Psychology, University of Stockholm, no. 27, 1979.

4 Occupational Health and Toxicology Branch, Department of Health, Visual display units, Wellington, New Zealand, 1979.
5 E. Gunarsson and I. Soderberg, Work with visual display terminals in newspaper offices - a visual ergonomic survey, The Swedish National Board of Occupational Safety and Health, AMMF 21 (1979).
6 The Swedish National Board of Occupational Safety and Health, Mental and social aspects of the occupational environment--general recommendations, 1980:14, Stockholm, 1980.
7 The Norwegian National Board of Occupational Safety and Health, Proposal to directions for visual display terminals, Oslo, 1981.
8 Action programme on computer policy for the Union of Civil Servants in Sweden, Stockholm, Sweden: Statistjanstemannaforgundel, 1982.
9 Electronic data processing in the social insurance office, Action programme for the Insurance Employee Union (In Swedish). Stockholm: Forsakringsanstaldas Forbund, 1981.

# OVERVIEW OF MANAGEMENT CONCERNS AND PROBLEMS IN OFFICE AND CLERICAL WORK

ANGUS G. S. MacLEOD
Center for Management Research and Education
Institute of Industrial Relations
University of California at Los Angeles
Los Angeles, California 90024

In 1927, the author E. B. White said, "I foresee a brilliant future in complexity in the United States." One way of depicting the complexity of life today, the impact of multiple changes resulting in a high degree of uncertainty and stress, is to visualize playing the fictional game of Chinese baseball. It is played exactly like American baseball except for one major difference. That is: After the ball leaves the pitcher's hand and as long as the ball is in the air, anyone can move any of the bases anywhere! In other words, everything in life is changing continually, not only the events, but also the rules governing judgments of those events. Often a change is perceived as a threat to security or to self-esteem, and results in anxiety and fear. In the ball game of life, everything is flux and all systems are open.

If you apply White's statement to the world of work and, in particular, to the field of occupational health, there have been a number of changes--technological, behavioral, etc.--which have made the field more difficult to comprehend and to manage than in years past. Today, management needs to be aware of labor force changes, for example, clerical workers are the largest sector of the white-collar work force (ref. 1), changes in life values, the expansion of government intervention as a result of labor legislation, the impact of technological change, research on new sources of stress and potential health hazards, the need for taking a systems approach on occupational health, costs incurred in dealing with job stress and occupational health, and the need for developing a management philosophy toward more effective utilization of human resources.

With the establishment of workers' compensation (WC) laws in the early 1900's, and for nearly 7 decades afterwards, attention was paid primarily to industrial safety and accident prevention and much less time and effort was focused on occupational health and industrial diseases.

The advent of OSHA early in the 1970's brought increased concern about occupational health and industrial diseases, for example, black lung, asbestosis, etc., which were still associated primarily with manufacturing or industrial work. Clerical and secretarial work, as contrasted to industrial or blue-collar work, was long considered as safe, clean, and healthy work.

Recent studies, however, show that white-collar work can be unsafe and unhealthy, as a result of polluted office air, stressful work pace, constant sitting in poorly designed chairs, etc. These things, which we once accepted as just part of the job, are now recognized as some of the numerous potential hazards to office and clerical workers.

Research (ref. 2) has shown that of 130 occupations, the second highest incidence of stress-related diseases is found in those who hold secretarial positions. A 1980 NIOSH study (ref. 3) showed that video display terminal (VDT) operators in strictly clerical jobs had the highest level of job stress of any occupational group studied. The 1980 Framingham study revealed that women clerical employees developed coronary heart disease (CHD) at almost twice the rate of other women workers (ref. 4).

These studies among others indicate to managers that occupational health hazards and the impact of stress on clerical workers are significant factors and are related to the effective utilization of human resources at work.

The purpose of this paper is to show (1) the increased costs that managements incur related to occupational stress and disease (with some discussion of the "cumulative injury" problem), (2) some causes of job stress among clerical workers, (3) some approaches to stress reduction (including discussion of employee assistance programs), and (4) some future developments that managements should consider.

## INCREASED COSTS INCURRED FROM STRESS

For a variety of reasons, including increased litigation, inflation, and increased government intervention, costs related to occupational health have risen sharply in recent years. Substantial increases have occurred in costs of workers' compensation, disability retirements, and health insurance. Suffice to say that such increased costs are a major motivating factor in terms of management interest and attention to occupational health.

One of the major cost elements is the great increase nationwide in cost of worker compensation coverage over the past 15 years. In 1965, the total was 1.8 billion. In 1980, 13.4 billion was paid out (personal communication, Research and Statistics Dept., U. S. Dept. of Health and Human Services, Social Security Administration, Washington, D. C.).

A large part of this increase is due to the growth of cumulative injury cases, sometimes referred to as "cumulative trauma." A cumulative injury case is one in which an individual has incurred a succession of slight injuries or traumas, none of which was disabling, but the cumulative effect of which has become disabling. A typical example would be a printer who eventually suffers a low back disability resulting from years of repeated bending and lifting.

A generation ago, interpretation and administration of the WC laws was much simpler than today. An injury or trauma (for example, a cut resulting from touching moving machinery), which could be traced to a particular incident and point of time was clearly compensable. However, compensability was highly questionable in the event of a disability that could not be traced directly to one given event.

Today, interpretation of WC statutes is much more complex and thousands of WC awards are made annually on the basis of a disability resulting from repetitive or "cumulative" trauma over a considerable period of time.

The following statistics on cumulative injury in California show a trend of rapidly increasing number of claims and substantial costs. In a recent 2-year period, cumulative injury cases increased approximately 22% for each of the 2 years. In a recent 6-year period, claims increased from 3.8% to 16% of all WC claims. The average cost of a cumulative injury case is more than four times the cost of an average WC claim.

In a recent 1-year period in California, a breakdown of types of cumulative injuries showed the following disabilities: 35% back injuries, 25% cardiovascular, 15% hearing, 13% extremities, 5% neuroses, and 7% all others. Many of these are definitely stress-related. Cardiovascular cases were the most expensive (ref. 5).

In regard to psychiatric claims, California and five other states have extended workers' compensation coverage to cases of emotional or mental illness caused by gradual or cumulative injury resulting from work stress.

For example, a secretary in Michigan won a $7,000 award for emotional distress caused by her boss's continual criticism and prying questions about her family life. A California court upheld an award to a legal secretary who claimed her breakdown was caused by the pressure of a heavy workload.

In Washington, D. C., the court allowed the claim of a clerical employee, who had worked for 20 years with no prior sign of emotional distress. This employee walked off the job and was hospitalized for a mental disorder that his attorney claimed was caused by the pressures of his job (ref. 6).

Another cost factor related to occupational health is the area of disability retirements. For example, in California, in a case that received much publicity, a public official was recently granted a disability retirement for a heart attack, which he claimed was caused primarily by harassment from another public official.

Costs of health insurance are also a major concern of management. Over the last 15 years, private health expenditures paid for by insurance have increased fivefold (ref. 7). Health insurance premiums have kept pace with this trend.

The increasing costs of job stress and occupational health are a major motivator for management interest in these fields. For managements interested in the cost effectiveness of measures to alleviate or prevent occupational stress, it is important to recognize and be aware of the causes of work-related stress.

## CAUSES OF JOB STRESS

The causes of occupational stress, no matter what occupation or level, are relatively similar. However, analysis to determine responsible stressors has become more complex. Although a single stressor may be readily identifiable, sources often interact with each other, and should be considered using a systems approach. Even though any one condition might not be stressful by itself, the combined impact of several conditions may lead to stress.

The major sources of office and clerical stress can be categorized as: (1) environmental, (2) job design, (3) employer-employee relationships, and (4) socioeconomic factors.

### Environmental sources

These include noise, lighting, inadequate ventilation, temperature and humidity problems, poor work station and furniture design, overcrowding, skin irritants.

### Job design

Some examples are rapid work pace or machine-paced operations, lack of control over time and work, constant sitting, monotonous, repetitive tasks, underutilization of skills, excessive overtime hours, pressure for higher productivity, servicing dissatisfied customers, work overload as well as underload, lack of adequate rest breaks, office automation, and status changes resulting from changing job requirements as a consequence of office automation. A good example of the last factor is the observation that assignment to a word processing machine (or a steno pool) may be very stressful to an accomplished secretary who has developed a close working relationship over many years with a given manager.

### Employer/employee relationship (including organizational "climate")

These include lack of respect, nonsupportive bosses, role ambiguity, role conflict, responsibility without commensurate authority, lack of participation in decision-making on matters related to one's job, overpromotion, lack of recognition for promotion, dead-end jobs, office politics, poor communication on employee problems, ineffective or nonexistent grievance procedures.

Socioeconomic factors

Some examples are low pay, job dissatisfaction, lack of child care, experience of discrimination based on sex, race, and/or age, increased workloads resulting from cutbacks, job insecurity (ref. 4, 8).

Study of such sources has become more complicated since they need to be viewed as interacting with extra-organizational sources, such as financial and family problems, life crises, and the psychological vulnerability of a given worker. A supervisor attempting to deal with an individual's stress may need to take several stressors into account. For instance, is it not likely that inadequate lighting and ventilation (environmental) could exacerbate an employee's feelings of dissatisfaction resulting from a monotonous, repetitive task (job design) on a dead-end job (employer-employee relations), which has low pay and job insecurity (socioeconomic)? The situation would become even more complicated if the employee's husband is dying of cancer and their son has had to leave college for economic reasons (extra-organizational).

If management is aware of cost increases related to occupational health and job stress for office workers, is sensitive to the impact of environmental and design factors, and is alert to the physiological, biological, psychosocial, and interpersonal aspects of technological and organizational change, all of which are major causes of work stress, then a rational and humanistic management will consider ways of (1) moderating and/or eliminating the effects of stress on workers and (2) assisting individual employees to cope with job stress.

## APPROACHES TO STRESS REDUCTION

A realistic approach to moderating or eliminating causes of stress would certainly encompass a restructuring of the technical and social environment, including increasing the social skills of supervisors. This mght be accomplished by taking a hard look at the four categories of work-related stress mentioned earlier and attempting to deal with stressors that are uncovered. In a sense, this action might be termed a stress audit, which could be done periodically.

Some of the things that organizations have done to assist individual employees to cope with stress on and off the job include disease prevention and "wellness" programs (nutrition, exercise, lifestyle changes, etc.), relaxation training, assertiveness training, and bio-feedback. (The Washington Business Group on Health, 605 Pennsylvania Avenue, S. E., Washington, D. C. can provide information on these programs.)

A rapidly spreading method of assisting employees to deal with mental and physical problems is the Employee Assistance Program (EAP). Several thousand companies, public institutions, and unions have EAP's. EAP has become the popular term for referral services that help employees whose persistent personal problems erode their job performance. It is also a generic term used to describe the various methods found in the workplace for the control of alcoholism, drug abuse, and mental health problems that adversely affect job performance. In the industrial setting, methods typically involve both rehabilitative and disciplinary components. Goals include the restoration of job performance to expected levels as well as the improvement of health.

The earliest EAP's (in the 1940's and 1950's) for the most part started as programs concerned with alcohol and/or drug abuse. Dealing with employees who have chronic drinking problems is still the major function of some EAP's. However, a large number of the more than 3,000 industrial EAP's in the United States, as well as the more than 35 EAP's in institutions of higher education, now provide counseling for employees with a broad spectrum of personal problems that interfere with job performance, including drinking and drug problems, marital, financial, job stress, parent-child difficulties, and other psychological problems. Such activities are referred to as "broad-brush" EA Programs (ref. 9, 10).

Some of the reasons why organizations install an EAP are: (1) need for reduction in cost of health services, (2) the fact that many physical ailments have a psychosomatic or emotional origin, (3) need to reduce excessive absenteeism, and (4) need to raise productivity.

Although some organizations believe that management should not "meddle" in the private lives of employees, and one authority refers to the spread of such employee counseling programs as "deadly paternalism" (ref 6), it appears that this type of program can be of major assistance as part of a stress reduction program. Although evaluation of EAP's is very limited, some major firms have determined that an EAP can be cost effective. Personal communications with the Equitable Life Assurance Society in New York City and the Kennecott Copper Company in Salt Lake City report a "return on investment" of five or six to one, primarily as a result of substantial reductions in absenteeism, disability income, and health insurance premiums.

Beyond using a stress audit to reduce or eliminate stressors, and helping employees to cope with stress, perhaps one of the most important elements in a stress reduction or prevention program is the philosophy of management in relation to the effective utilization of human resources. Management's views on upward mobility, career development, planning of change, social support systems, training supervisors in social skills, the extent to which employees exercise autonomy over their work, and the degree of participation in decision-making are critical factors in developing an "organizational climate" in which many work stressors are modified or eliminated.

As far as future developments are concerned, we can hope for further expansion of these activities related to effective utilization of human resources.

SOME FUTURE MANAGEMENT CONCERNS REGARDING JOB STRESS

There are three areas or questions related to occupational health that managements may need to consider more fully over the next several years. One is the question of corporate responsibility for employee mental health. In a pending court case, a widow is suing her husband's employer for $6 million, claiming that the company caused her husband's death by suicide when it failed to respond to his repeated complaints of overwork. At issue is the extent to which a company may be held responsible, beyond the provisions of the workers' compensation laws, for a psychiatric injury to an employee (ref. 6). The outcome of this case bears watching since it highlights a growing concern and interest in work stress and also because it is a part of two larger questions.

Those questions are: What is management's responsibility for the overall health of their employees?" and "Can managements play a role in improving the level of health services on a national scale?" Relative to these questions, it is worth noting that a new occupational category has appeared recently in industry--the health service professional or health services manager. Some of these persons, who are not medical doctors, run EAP's and are also heavily involved with health education and disease prevention training programs in their organizations.

In this context, one may also wonder whether, since unions are increasingly interested in work stress, the development of joint labor-management committees on stress reduction would be a feasible and useful development. The well known adversary relationship between labor and management may be a stumbling block, but it is possible that such joint activity would be advantageous to management, union, and all employees.

In summary, E. B. White's prediction of a "brilliant future in complexity" certainly applies to occupational health and work stress. The role of chronic job stress in physical illness and psychological strain is indeed complex. Physical and emotional responses to stress are not separable but build on one another. There is enough evidence from research findings to indicate causes of stress and steps that can be taken in reducing and/or eliminating stress from office workers' jobs. A cost-conscious and humanistic management may well find that stress reduction activities pay substantial dividends in cutting costs and increasing worker productivity.

REFERENCES


1 Bureau of Labor Statistics, Employment and earnings--annual averages by occupation and industry--U. S. Labor Force, U. S. Department of Labor, Washington, D. C., 1981.

2 M. J. Smith, M. J. Colligan and J. Hurrell, Jr., A review of NIOSH psychological stress research, proceedings of a November 1977 Conference on Occupational Stress, co-sponsored by NIOSH and the UCLA Institute of Industrial Relations, pp. 27-28.

3 M. J. Smith, B. G. F. Cohen, L. W. Stammerjohn, A. Happ, An investigation of health complaints and job stress in video display operations, Human Factors, 23, 1981, pp. 387-400.

4 Health hazards for office workers--An overview of problems and solutions in occupational health in the office, Working Women Education Fund, Cleveland, Ohio, 1981.

5 Cumulative injury in California--A report to the industry, California Workers' Compensation Institute, San Francisco, California, 1977,

6 B. Rice, Can Companies Kill?, Psychology Today, June, 1981, 81.

7 Health Care Financing Administration, Health Care Financing Review, U. S. Department of Health and Human Services, winter 1980.

8 C. L. Cooper and J. Marshall, Occupational sources of stress, in Proceedings of a Conference on Occupational Stress, co-sponsored by NIOSH and the UCLA Institute of Industrial Relations, 1977.

9 J. T. Wrich, The Employee Assistance Program, Hazelden Literature, Center City, Minnesota, 1974 and 1980.

10 M. Shain and J. Groeneveld, Employee Assistance Programs, Lexington Books, Lexington, Massachusetts, 1980.

# COLLECTIVE BARGAINING AND OSHA AND CLERICAL WORKERS

JUNE McMAHON
Director of Research
Service Employees International Union
2020 K Street, N. W.
Washington, D. C. 20006

## UNION MEMBERSHIP IS BROAD BASED

Occupational safety and health has always been a top priority item in negotiating for Service Employees International Union (SEIU) locals. Historically, SEIU has been known as the "janitor's union," but it was the strength of its bargaining practices and the raises and protection locals won, that influenced other groups of workers to seek membership. This resulted in a name change in 1968 from Building Service Employees Union to Service Employees International Union to better reflect the true extent and diversity of the membership. Most of SEIU's membership is concentrated in hospitals and nursing homes, the public sector--state and local government and building services, including schools. Every service occupation is represented, including sellers, secretaries, nurses, lawyers, clerks, ushers, and technicians. Clerical and white-collar membership cuts across all industry lines. Our membership includes ward clerks and office clericals in hospitals and nursing homes, typists, stenographers, office machine operators, and hundreds of clerical and office worker classifications in the public sector, including clerks and secretaries in public schools. Recently, we entered into a campaign in conjunction with Working Women to bring trade unionism to the country's more than 20 million secretarial and clerical workers, one of the biggest but least unionized segments of the U. S. work force.

## GROWING AWARENESS OF WORKPLACE HAZARDS

Clerical workers are beset on all sides by chemicals and hazards that threaten their well-being and their ability to do their jobs. Accidents and illnesses caused by tripping, poor ventilation and sealed buildings, chemicals used in photocopiers and office equipment, noise, excessive sitting or standing, poor lighting, VDT's, monitored communication equipment, and stress all affect the health of these workers. For so long, the headaches, irritability, fast heart beat and breathing, and stomach problems were chalked up to "women's problems" or a high strung personality. In SEIU, our clerical workers were faced with a few more hazards. School clericals found their workplaces surrounded by asbestos. Sometimes, the workers are assaulted by ill-behaved children and receive bruises, cuts, and bites. Hospital clericals in understaffed hospitals and nursing homes were asked to take on aide duties, which compounded their stress and their workloads. Clericals in the public sector are faced with the uncertainties of lay-off and unemployment that accompany the fiscal crisis of our cities and states. Our local in Contra Costa County, California, forced the evacuation of the County Social Services building after it was determined that the facility was constructed on an abandoned chemical dump. While the concentration of strong chemicals was below state standards, little is known about the long-range effects of the chemical exposure.

In 1980, Jeanne Stellman from the Women's Occupational Health Resource Center started writing a column for our newspaper, Service Employee, on the occupational safety and health problems that face our industries.

We did a breakdown of the responses to the column in order to better direct our safety program. The column that elicited the most response, either requesting further information or citing a problem for the writer, union, or co-worker was on reproductive hazards in the workplace. The second highest response was to the fire hazards of plastics and the dangers of toxic fumes in a sealed environment. The illnesses and injuries that result from these hazards frequently befall women clerical workers.

Scrutiny by industrial hygienists and research by safety and health experts has helped to publicize the problems of workers in all industries. Research results have indicated that workers face a multitude of hazards--frequently not discernible to the untrained eye.

On every count, employers have failed to provide adequate protection for workers. Voluntary efforts by employers have proved to be insufficient or nonexistent, and it has only been through collective bargaining and the passage of OSHA that workers have gained the most in terms of a right to a safe and healthy workplace.

## UNION APPROACH TO REDUCING HAZARDS

In the past, SEIU has endeavored to train its members in the identification of hazards in the workplace. In conjunction with the University of Wisconsin and the University of Illinois, we developed a program to help workers identify and correct the hazards found in hospitals. Hundreds of SEIU members have passed the course, but most of our emphasis has been on the development and bargaining of strong safety and health language in our collective bargaining agreements. Strong, clear language that gives workers the right to refuse work they consider hazardous has been our goal. It has been difficult to achieve in most cases, and faced with the administrative weaknesses of OSHA, the workers' best recourse has been the grievance procedure that is available under a collective bargaining agreement. Both the general safety and health clause and the grievance clause have proved to be a worker's best protection against a hazardous workplace. The simplest general contract language provides for a safe and healthful workplace and the employer agrees to be in compliance with recognized safety and health standards. This type of language is the most common in SEIU agreements, and we are working on provisions that go beyond this--for example, the right of a worker to remove herself or refuse to do work she considers hazardous. This is one of the most sensitive subjects at the bargaining table, as it hints of moving in on the sacred ground of "management rights." SEIU negotiators are making demands for some form of guarantee that their members will not be penalized for not performing work under hazardous conditions. While a worker is somewhat protected by OSHA under Section 502 of the Labor Management Relations Act, which provides that "quitting of labor by employees in good faith because of abnormally dangerous conditions for work" is not a strike under the Act. A worker is not guaranteed pay for the period during which services are withdrawn because of "abnormally dangerous conditions."

Most arbitrators agree that under proper conditions an employee may refuse to obey an order to work in a hazardous situation, however disciplinary action may be upheld where (1) the dangerous aspects of the job have been eliminated by the employer, (2) the employee's job necessarily involves hazardous work, or the employer has already factually determined that the work is not dangerous. These may be narrowly defined limitations, but negotiators have pushed for clauses indicating that the individual employee has a right to immediate relief from the assignment without penalty.

A number of SEIU contracts provide for immediate joint investigation upon complaint by an employee that the work is unsafe, and for the handling of the problem through an accelerated grievance procedure. Additional language is negotiated into the agreement guaranteeing the individual the right to another assignment without loss of seniority, status, or pay.

In many cases, the results of research in occupational safety and health have provided the factual basis for the refusal of unsafe work clauses. If the local can show data measuring the severity of these types of hazards, part of the job at the bargaining table is done. Unfortunately, the documentation of hazards in so-called "safe" jobs in hospitals and office buildings has not been a high priority. Slowly, research in the field is coming into its own.

## WORK-RELATED STRESSES AFFECT FAMILIES

The labor movement is finally beginning to realize how the health and safety of a workplace affects not just the individual worker but the worker's family. In the past, we have negotiated health insurance, dental care, vision care, and educational benefits for the spouses and dependents of our members, but it has taken the dangers of the chemical and stress-related hazards in the workplace that we bring home to our children, our unborn children, and our spouses to awaken us to the fact that the collective bargaining process must have a broader mandate. It is time for unions to present a program for workers and families.

Consider the world of work and beyond. Women today continue to have primary responsibility for childrearing and homemaking. Most women workers, and especially clerical workers, face jobs that are increasingly stressful; the worker's satisfaction with the job decreases, and the ability to share the warm satisfying aspects of the job with family and friends is diminished. The fragmentation of work, and the deskilling of the work force, combine to create conditions in which the individual worker experiences stress. Stress is often interpreted as a personal problem instead of a problem that could be solved by collective action. It is even more difficult to identify the problem, since stress manifests itself differently for different workers--as headaches, neck or back tensions, high blood pressure, colitis, insomnia, depression, withdrawal, alcoholism, drug abuse, or frantic activity. Most women rarely understand that they are facing a common work-related problem. Instead they feel negative about themselves for having stress symptoms and these feelings have a direct impact on their family lives.

Typically, stress is not handled through collective bargaining to change the situation but by the worker denying the cause of the stress and internalizing the stress. It is then brought home, manifesting itself in tensions and irritations that grow out of hand.

This is only one example of the work-related problems that workers, especially working women, face in their personal lives in the key intimate relationships with spouse or children. When a woman finds herself expressing anger to a child or loved one that seems out of proportion to the immediate issue, very often what is really happening is a release of the tensions and frustrations that have been brought home from work.

What can be done through collective bargaining? First, a good health and safety clause with prompt enforcement and recourse under the grievance procedure, and second, a freedom of choice. Clauses can be designed to allow workers some freedom of choice over their work. Examples include the worker's right to immediately refuse work she considers hazardous, freedom to choose flexible hours, rest periods, and work weeks, and the availability of optional education and training programs.

None of these measures substitutes for the provision that the employer must provide a safe and healthy workplace. But coupled with family needs, the collective bargaining agreement can truly be an instrument that provides not only for a better working life, but for a better life in general.

# THE VIDEO DISPLAY TERMINAL (VDT) USER'S PERSPECTIVE

MARINE W. McGEE
Secretary, U. S. Attorney
220 Post Office Building
Cincinnati, Ohio 45202

Working various clerical and secretarial positions as clerk-typist for the U. S. Department of Agriculture, the Division of Air Pollution, Environmental Protection Agency, Solid Waste Information, and the Centralized Services Branch, Internal Revenue Service; as a Medical and Psychiatric Secretary, Jewish Hospital, Medical Records; Secretary, Research Group, NIOSH, Secretary, Regional Office IRS, Fiscal Management Branch, Interim Secretary District Office IRS, Training Section, and currently, Secretary (VDT operator) U. S. Attorney's Office, I feel I have encountered many aspects of job stress.

These positions are mentioned to give an idea of the types of pressure I have experienced through the years, including oversupervision, dead-end positions, unclear and conflicting job responsibilities, work overload, work underload, inflexible time schedules, constant criticism, and, sometimes, just plain boredom.

I can remember that when I first began in the Federal service I had a job typing government bills of lading for nonfat dry milk 8 hours a day, 40 hours a week with two 15-minute breaks and a 30-minute lunch break; a total of 1 hour away from the typewriter each day. I actually dreamed about these things when I retired at night. On one occasion, my husband, who worked the 3 to 11 shift, came home and asked for his nightly snack. I told him to look in the top basket on the right side of my desk. He continued to ask me and I continued to tell him to look in the top basket. He woke me up because he was laughing so hard. I realized then I was under considerable stress. I literally worked the job, slept the job, and was trying to feed my husband the job. You can imagine how delighted I was some months later to learn the job was leaving for another city.

Turning from my past, I'd like to discuss my current position as a video display terminal operator with the U. S. Attorney's Office. I have found the VDT a new concept in the work field. I am learning to use it with some proficiency now and greatly enjoy my work on the equipment. I use the IBM System 6 Word Processor, which transmits, retrieves, stores, and edits information. In our office, I am referred to as the "dedicated System 6 operator," using the equipment on a full-time basis to perform tasks for a variety of lawyers and secretaries. I type from hand-written drafts and cassette tapes of letters, memos, and exhibits lists for trial preparation. The legal documents are retrieved by a "recall" command, where I scan to variables and fill in the blanks.

By contrast, some of the secretaries in my office use the second VDT machine on a part-time basis and frequently ask me about the classic symptoms described by the experts, such as headaches, blurred vision, eye strain, nausea, fatigue, and sore necks. Most of my co-workers who are plagued with these complaints have developed a dislike for the equipment. Often they ask, "How can you sit there all day and use that machine without getting a backache." Fortunately, I have not suffered from the customary VDT problems. What's more, I do not have the pressures of my previous secretarial jobs and I like what I'm doing. In my opinion, my attitude toward my work plays an important role in my being able to sit at the machine without any complaints.

The recent studies indicate that VDT operators have one of the most stressful jobs in our workforce today. A local newspaper carried an article on this subject a few weeks ago and the following morning I was greeted by one of our staff with the fact that I should "be one of the most stressed in the office."

Being in the secretarial field, most of us suffer from what I call the type and retype syndrome. This syndrome can cause a lot of stress-filled hours. Many secretaries have had what was thought to be a final copy, ready to be mailed or sent to the publisher, when suddenly it reappeared in the typing pile to change "was" to "were" or "and" to "but" so it had to be retyped. Probably, when it was first typed, the present tense was correct but by the time it was typed and retyped, the past tense was more appropriate. During the retyping there is always the risk of making a typing error; by using the VDT or some other word processor, however, there is no need to retype the entire page, just call up the page number, make the change, and rerun the corrected copy.

As a matter of fact, a few months before I left one of my jobs where there was plenty of typing and retyping with no VDT available, while on vacation I developed a dull, nagging pain in my left shoulder. This constant pain lasted for approximately 9 months, at which time I decided to change jobs. I believe the pain represented subconscious fear of returning to all the pressures of the job. After about a month on the new job, I noticed one day the pain had disappeared. To be truthful, I don't know when or how it left, but my new surrounding was less demanding and pressures were reduced to a minimum, so the pain disappeared.

I cannot attest to the findings of the researchers, since I find my job relaxing and enjoyable. My workload, although heavy at times, does not have the constant deadlines. By planning ahead on the work, I can alleviate most of the last-minute rush. My equipment is ideally situated away from the outside light, there is no particular strain on my eyes, and my chair position is comfortable. Last, but certainly not least, I have developed a positive attitude toward the equipment itself. I find it exciting to be able to use the machine correctly, and it performs as expected. In short, I enjoy using the equipment.

In conclusion, I'd like to mention a few quick tips that I employ to reduce potentially harmful stress. I find they apply to the particular stresses of my job as a VDT operator.

1. When you know you are going to have a stressful situation, think about it and decide what action you will take to deal with it beforehand. If two attorneys want something at the same time, don't become angry. Discuss the problem with them. Find out which job has priority and do that one first. Keep the lines of communication open.

2. Learn to be patient. Pace yourself to respect the slowness of human and mechanical processes. Instead of getting upset if the machine jams, call the repair service and take a much-needed break.

3. Take control of your leisure time. Make plans to visit some place you've wanted to see for a long time and just go, whether it's the art museum in your city or the Statue of Liberty in New York. When you return you will be rested for your job and more enthusiastic about doing it.

4. Adjust your behavior so that it is appropriate to the setting. It's okay to be aggressive and tough at the office. At home, it's usually better to use TLC (tender loving care).

5. Slow down on worrying, 95% of what you worry about doesn't happen, and the rest you can't do anything about anyway.

6. Try not to offer unsolicited advice to adults. It will probably make them angry and cause you both unnecessary stress.

# CLERICAL HEALTH AND SAFETY: STRATEGIES AND ACTIVITIES BY PUBLIC SECTOR UNIONS

SUSAN SILBER
Gagliardo & Silber
1101 Connecticut Ave., NW
Suite 409
Washington, DC 20036

KATHERINE COLGAN
National Labor Relations Board
Washington, D. C.

## INTRODUCTION

This paper focuses on the Federal sector work force partly because it is what distinguishes our perspective, but also because the United States government is the nation's single largest employer, with a correspondingly massive number of clerical workers, and because the Federal clerical work force is unionized beyond any comparable group of office workers.

Even though unions have exclusive rights to represent a significant number of clericals does not mean that clerical issues are sufficiently prioritized by these unions, that many secretaries comprise their dues-paying memberships, or that they are represented in the leadership of the Federal unions. For example, one typical Federal union that claims 50% of its members are women has not one woman (and certainly no clerical worker) on its national executive board.

In harsh social and economic times such as these, all Federal unions are struggling not just to preserve the status quo but to minimize the losses brought on by massive lay-offs (called reductions-in-force in the Federal sector), parallel speed-ups for the workers who remain, and the budget-cutting ax to many hard-won fringe benefits. Increased automation and heightened production quotas prevail. Given this defensive position, it is understandable that unions might not be taking the initiative sufficiently on invisible issues such as clerical health.

Unions have the potential to contribute to the movement for health rights and are the predominant means of achieving and institutionalizing employees' striving for healthier worksites. Struggling for healthier working conditions involves more office workers in their unions and thus empowers both. In the Federal sector, where unions cannot bargain over bread and butter, unions must focus on head and heart. These demands are backed by a rich tradition of women's activities in the trade union movement. In the 19th century, these women sang "Give us bread, but give us roses, too."

It is now, just when the working conditions in the Federal sector are worsening, that a fight for clerical rights is most needed. The Federal resistance. For example, an Internal Revenue Service district office recently shortened the lunch period allegedly to improve productivity--but only the lunch period of the clerical staff. The policy was reversed after demonstrations and publicity spurred intense negotiations. The district director had believed that secretaries "normally" get less time for lunch and he could get away with the change. The union pointed out that secretaries suffer the same indigestion as other workers and deserve equal working conditions. Not only did the workers win back their previous lunch period, but the union also greatly increased its membership.

## LEGISLATIVE AND REGULATORY CONTROLS IN THE FEDERAL SECTOR

Collective Bargaining in the Federal sector is governed by the Civil Service Reform Act (CSRA) (ref. 1), enacted in 1978. Title 7 of CSRA affords parallel protections to Federal employees as are guaranteed private sector employees under the National Labor Relations Act (ref. 2). These include the right to organize and the right to bargain collectively--with certain limitations to be discussed *infra*.

## LEGISLATION AND REGULATION CONCERNING HEALTH AND SAFETY

Section 19 of the Occupational Health and Safety Act of 1970 (ref. 3) directs each Federal agency to establish a comprehensive health and safety program consistent with standards applicable to the private sector in Section 6. The statute requires the Secretary of Labor to evaluate the effectiveness of each program annually and to report to the President. Title 5 of the United States Code (ref. 4) also directs agency heads to establish programs and further requires that accidents and injuries be reported to the Secretary of Labor.

In 1980, President Carter issued an executive order calling for further implementation of OSHA Section 19 (ref. 5). The order (1) directs the Secretary of Labor to issue standards for agency programs, called basic program elements, and (2) defines the scope of responsibility of several entities including agency heads, joint health and safety committees, the Secretary of Labor, the General Services Administration, and Field Advisory Councils on Occupational Health and Safety. The order further provides that official time shall be accorded all workers while engaged in activities under its terms.

## THE DEPARTMENT OF LABOR REGULATIONS

On October 21, 1980, the Secretary of Labor promulgated regulations (ref. 6) called Basic Program Elements for Federal Employee Occupational Safety and Health Programs. These cover all Federal employees, including volunteers and prisoners (ref. 6). The stated purpose of the regulations is "to assure safe and healthful working conditions for Federal employees" (ref. 7).

Agency health and safety programs

Agency heads are primarily responsible for establishing programs in their own agencies that comply with the regulations (ref. 7). This includes budgeting for personnel, testing, training, medical surveillance, and abatement of hazards. The agency head must also determine if hazards exist in his agency that are not covered by OSHA standards. If they do, he must adopt supplemental standards where necessary and appropriate (ref. 7).

Employees' rights

The regulations afford employees certain rights.

(i) Right to report. Federal employees have the right to report unsafe and unhealthful working conditions to appropriate officials without reprisal. They may request that their names be withheld. Time periods keyed to degree of risk are established within which an OSHA inspection must be conducted. If no inspection is ordered, the employee is entitled to a report of the reasons in writing, with a copy to the certified committee, if there is one (ref. 7).

(ii) To accompany inspection. Employee representatives shall be given the opportunity to accompany inspections. Those who do so must be afforded training in hazard recognition and evaluation appropriate to the work involved. Several examples are listed, of which clerical work is first (ref. 7).

(iii) To seek NIOSH assistance. Where there is no certified committee, Federal employees may request NIOSH to determine whether or not hazardous conditions exist at their workplace (ref. 7).

(iv) To refuse to work. Federal employees may refuse to work if they reasonably believe that they are in imminent danger and have insufficient time to seek redress through normal procedures (ref. 7).

(v) Official Time. Employees have the right to participate in covered activities on official time. In the Federal sector, official time is construed to include travel and *per diem* expenses. With regard to attendance at Field Council meetings, the regulation specifies that travel funds shall be available equally to management and nonmanagement employees (ref. 7).

Enforcement and abatement

Annual OSHA inspections are required at each worksite, including office operations. Reports by employees and private interviews with employees are encouraged during inspections (ref. 7).

Where a hazard is found, the inspector issues a Notice of Unsafe or Unhealthful working conditions, describing the harm and fixing a period for abatement. This notice must be prominently posted by the agency health official and a copy sent to the certified committee or employee representative (ref. 7). Unannounced follow-up inspections may be conducted until the condition is cured (ref. 7).

General Services Administration (GSA)

The regulations direct the Secretary of Labor and the Administrator of the GSA to work cooperatively to resolve conflicts in their standards. The GSA is specifically charged with providing space meeting the health and safety requirements of each agency and with providing safe equipment and supplies. Regarding space, the GSA must cooperate where necessary with occupant agencies to abate hazards. Regarding materials, the GSA must initiate corrective action when an agency reports items purchased by it to be unhealthful. Information from Material Safety Data Sheets, which are used when purchasing hazardous substances, is maintained on line. GSA representatives must accompany OSHA inspectors upon request (ref. 7).

Joint health and safety committees

An agency may establish a joint committee comprised equally of employee and management representatives whose purpose is to "monitor and assist" the agency's program. For agencies in the field offices, two types of committees are required, one at the national level and others in the local offices. Agencies must furnish committee members with appropriate information and training (ref. 7).

Whether or not the agency has such a certified committee affects the scheduling of OSHA inspections. When there is no committee, OSHA may conduct unannounced inspections. Where a committee has been established, OSHA may inspect only in circumstances (1) where half the members of the committee request the inspection, and (2) where an employee has reported an imminent danger and neither the agency nor the committee has satisfactorily responded (ref. 7).

The committee may also report dissatisfaction to the Department of Labor, if one-half of the membership finds deficiencies in the agency's program or its investigation of allegations of reprisal (ref. 7).

Field Federal Safety and Health Councils

Field Federal Safety and Health Councils have been established in regions where there are significant concentrations of Federal workers. Their purpose is to "facilitate the exchange of ideas and information throughout the government about occupational safety and health (ref. 7). The regulations provide for quarterly meetings, selection of employee representatives, and official time and travel fees.

Secretary of Labor

It is the responsibility of the Secretary of Labor to facilitate the effectiveness of all the provisions of the regulations. Among other things, this includes providing appropriate training, furnishing technical services and educational materials, evaluating agency programs, maintaining agency programs, and maintaining records, including a log of Federal occupational illnesses and injuries (ref. 7).

## PROBLEMS OF FEDERAL SECTOR UNIONS

Collective bargaining impediments

The major impediment to bargaining in the Federal sector is the deprivation of the right to strike, the union's ultimate and most effective economic weapon. Federal sector unions are correspondingly more dependent upon labor boards, arbitrations, litigation, and legislation than comparable private sector unions would be. In the place of strikes, the Federal Services Impasses Panel provides a forum for binding arbitration when the parties reach an impasse in negotiations (ref. 8).

Basic bargaining rights are limited in two other important ways. First, the right to organize does not include the right to a union shop although the union has the duty to represent those who have chosen not to join. Hence, Federal sector unions are often hampered. Second, the right to bargain concerning terms and conditions of employment is very narrowly limited by broad management rights statutory language (ref. 9).

Health and safety problems in the Federal sector

OSHA standards present two types of problems. First, they are designed to assess risk where a single hazard is predominant. It is difficult to apply them where multiple factors create an unhealthy environment. This problem was faced by workers at the Brookhaven IRS Service Center.

> For years, employees complained of burning eyes, sore throats, sinus problems, vomiting, dizziness, headaches and a metallic taste. NIOSH investigators found some evidence that diesel fuel contaminants enter the buildings via rooftop air vents, but "no conclusive toxicity" explaining the symptoms. They point out that the problems are "particularly frequent and distressing" in the Data Conversion, Collection and Computer Room areas, which they describe as follows: "Many of the employees in this building are working under conditions of considerable stress. In the data collection and conversion areas, workers are required to enter tax info onto microfilm machines using VDT's. (They are under pressure to process a large number of forms,

their productivity is measured and they are evaluated accordingly . . . (ref. 10).

The NIOSH investigators conclude that: "These additional stress factors clearly may act to potentiate any health effects from fumes, and should be considered when planning measures to correct the problems."(ref. 10)

Certain problems also arise in implementing joint committees on the OSHA model. Membership in the committees is supposed to be 50% worker representatives and 50% management. The employee representatives are appointed by the agency, either from a list submitted by the union, or, where there is no union, by a process the agency develops itself. When some workers are organized and others are not, committee membership is supposed to represent both groups (ref. 7). Thus, where substantial numbers of employees are not organized, it is very possible that union representatives will constitute less than 50% of the committee.

At a worksite where the union is already weakened by an open shop, this could be a serious problem. Where low union consciousness results in ineffective committee representation, management may be permitted to remain inactive and OSHA may never be called in (ref. 7).

There is a controversy as to whether or not it is better to have a committee where the union members are outnumbered. Some argue that it is better to have a union committee and avoid the restriction on inspections. Others argue that where a certified committee exists, management has the obligation to meet with employees whereas it may ignore a union committee. In addition, inspections may still be ordered following direct employee reports to OSHA where the committee has not responded to an employee's concern (ref. 7).

Finally, OSHA under Section 19 has no effective enforcement power. Its most stringent sanction is the power to notify an agency that a hazardous condition exists and to request abatement. Therefore, in order to obtain the protection guaranteed by statute, by executive order, and by regulation, Federal sector unions find it necessary to bargain for them.

ACTION BY FEDERAL SECTOR UNIONS

Federal sector unions are addressing the problem of safe and healthful working conditions for office workers primarily in two ways: employee education and contract language. The three unions to be discussed are: American Federation of Government Employees (AFGE), National Federation of Federal Employees (NFFE), and National Treasury Employees Union (NTEU).

Employee Education

There are three components to this area: research, literature, and training, with different degrees of emphasis in each union.

(i) Research. The NFFE has a Legal and Labor Relations Department with a strong emphasis upon investigating the types of hazards faced by office workers such as stress or poor ventilation, and developing specific means of resolving them. For instance, solving problems of VDT workers involves considerations of furniture design, paint colors, and regular equipment inspection, as well as *de facto* eye tests and frequent breaks. Another union, the AFGE, the largest of the three, has an industrial hygienist on staff.

A study of great interest to public sector unions is currently being conducted by Drs. Jean Stellman and Gloria Gordon of the Columbia University School of Public Health. The study will measure the health and well-being of office workers as affected by office type, environmental factors, and job characteristics. The participants in the study are public employees who belong to unions, a new data base particularly applicable to problems of public sector unions will thus be created. It is the intent of the investigators to make their findings available directly to unions.

(ii) Literature. Two unions are currently preparing literature and training material for the membership. One technique that has been effective in raising employee consciousness is the use of questionnaires. AFSCME has prepared a few such questionnaires. Another possibility is an OSHA inspection checklist.

(iii) Training. AFGE places its emphasis upon training. The union is developing a comprehensive training program for all types of employees under a 3-year model training grant from OSHA. The union is divided into 15 districts nationwide, each of which has a Safety Coordinator. The Safety Coordinators attend a national training program given at the University of Wisconsin. Each district also has a local training program for shop stewards and union officers conducted by a specialist from union headquarters.

As part of the program, a workbook and a manual for local use were developed and circulated. The materials are very practical and focus on building chapter strength through resolution of health and safety issues.

### Contract language

Two unions, AFGE and NFFE, have prepared model contract language concerning clerical issues. Both adopt the joint committee approach. There is further language concerning: adequate light, ventilation, and space (NFFE); physical examinations for those exposed to hazardous or unhealthful working conditions (AFGE); periodic screening for contaminants (NFFE); facilities for providing prompt medical treatment (NFFE); and removal without loss of pay or leave of employees from areas where temperature or humidity are unacceptable (AFGE).

Which of these benefits are actually won depends upon the strength of the bargaining unit and the importance of health and safety issues in relation to other problems. The NTEU in its last contract with the Internal Revenue Service won a general duty clause cast in positive terms, a Safety Advisory Committee with official time for meetings and inspections, a cancer detection program, and a commitment from management to make efforts to reassign workers grieving health and safety issues pending the outcome of the hearing.

The importance of a general duty clause in the contract is based on the fact that OSHA standards are commonly regarded as ceilings by Federal supervisors. Workers at the Voice of America recently won an arbitration confronting this issue. They had complained repeatedly of dust and odor emitted from a certain brand of recording tape provided by the GSA. The problems were aggravated by the workers' confinement to poorly ventilated recording booths. The agency claimed that no OSHA standards had been violated and that therefore no remedial action was necessary. The two unions involved, AFGE and NFFE, jointly submitted the dispute to the State Department Foreign Service Grievance Board. The Board found unhealthful conditions to exist in violation of the collective bargaining agreement, irrespective of OSHA standards. Pursuant to the Board's order, the agency now uses a brand of tape acceptable to the employees. The agency was also directed to negotiate with the GSA regarding changing specifications for recording tape, and to obtain material safety data sheets from suppliers.

Federal sector unions find it necessary to attempt to incorporate certain OSHA basic program elements into the collective bargaining agreement. Troublesome areas are getting an acceptable general duty clause and recognition by management that the agency, not the GSA, has primary responsibility for employee health and safety. It is necessary to convince the agency that it would be harder to fight the union than it would be to fight the GSA for changed specifications, exemptions for using GSA materials, and procurement of and access to material safety data sheets. Employees currently have great difficulty obtaining product ingredient information on the less dramatic substances such as "white-out."

A second important area to consider in the contract is formation of committees and conduct of inspections. Formulation, composition, purpose, and administrative support should be spelled out. So should time limits for inspections, union recipients of inspection reports, and appropriate training for inspectors and committee members. Although very clearly stated in the Executive Order, official time with travel and per diem expenses should be guaranteed. This is particularly important with reference to Field Federal Safety and Health Councils, which have been attended primarily by management to date. Employees should be encouraged to attend Field Councils to make the union position felt at the interagency level.

## BARGAINING STRATEGIES

There are a number of traditional approaches available in coping with health and safety issues. The introduction of VDTs is a good example because several types of challenges can be used. First, the introduction of new machines is a change in working conditions. When working conditions are changed, negotiations on the impact and implementation of the changes are mandated by Federal law. Health impacts are no exception. Thus, even when the change itself is nonnegotiable (such as the institution of new technology), the union can best reach the table by framing proposals to mitigate the adverse effects of the change on the health of the workers.

For example, when an Internal Revenue Service district office proposed to convert to "open space," NTEU submitted proposals concerning the installation of partitions, drapes, white noise, acoustical carpeting, and individual desk lamps. The agency refused to negotiate, claiming nonnegotiability based on management's right to select technology (ref. 11). NTEU submitted a negotiability appeal to the Federal Labor Relations Authority arguing that it should not be foreclosed from bargaining as its proposals do not prevent management from going to open space, but simply provide that such an arrangement be implemented in accordance with standards of health. [The Authority did order the agency to maintain the status quo and bargain with the union. In the wake of the decision, the agency decided not to convert to open space.] This illustrates another problem with the Federal sector collective bargaining--long delays for the resolution of negotiability disputes. However, if the union is eventually successful, the standard remedy is a return to the _status quo_ before the changes and a resumption of bargaining.

Where VDTs are hazardous in themselves or are used under adverse circumstances, such use may be challenged under the health and safety clause. It is particularly important for Federal sector unions to have strong general duty clauses. Also, attempts should be made to include health and safety limitations in every relevant area of the contract. Some possibilities are considered below.

### _Equipment and materials_

Some unions have language to the effect that production quotas may not be set so as to affect the health and safety of employees. Similar clauses are appropriate in language referring to maintenance of equipment and specifications for materials. Depending upon the nature of the hazard faced, the contract might call for relevant physical examinations (Newspaper Guild); appropriate furniture, lighting and paint; or periodic radiation testing.

Tours of duty and workloads.

Tours of duty are also important to employee well-being. Frequent break times, flexible scheduling, and rotation of duties may prevent certain recognized occupational illnesses. Terms concerning maximum workloads or case loads may reduce overwork and stress. Exercise programs should be made available.

Job classifications.

Downgrading of job classifications resulting in less control over jobs or the introduction of piecework, should be strenuously avoided as creating stress and having an adverse effect upon health. The corrective to this is the development of career paths and training and upgrading programs so that employees may learn new technology. Unions should be beginning now to bargain with a view toward preventing undesirable changes in job structure and to increase affirmatively the involvement of workers in the selection of technology and redesign of work. Management must be reminded of the health-related imperative of these demands, the greatest danger in jobs with high demands but low control (ref. 13).

Medical treatment.

Treatment of occupational illnesses and injuries is also a bargaining point. Illnesses relevant to the workplace should be recognized in the contract. The union should also seek to shift from a Workers' Compensation approach to a contract approach by incorporating treatment and prevention programs into the contract itself. But unions also should file workers' compensation claims based on stress.

Disciplinary procedures and job security.

It is important that employees who leave a job because of imminent danger should not be disciplined. The union should seek an agreement that there be a hearing prior to any suspension or discharge for such a reason. General procedures to protect employee job security alleviate the worst stress of all--the threat of job loss. Unions successfully use health components such as improper lighting, sexual harassment, stress, and employer unfairness as defenses in discipline cases alleging unsatisfactory performance.

Grievance procedure.

Where workers grieve health and safety issues, management should transfer the grievants without loss of pay pending the outcome of the hearing. Such a transfer should not moot the grievance. Because of the seriousness of health and safety issues, an expedited grievance mechanism should be sought.

Other contract issues. Other contract issues that alleviate stress, especially for women workers, should be pursued such as: a strong antidiscrimination policy; prohibition of sexual harassment, confronting pay inequities and underevaluation of work; and provisions for child care.

COALITION BUILDING

Another methodology for forming a strong program in pursuit of healthful working conditions for the clerical work force is through coalition with other groups. One model in the Federal sector is the Federal task force of the National Committee on Pay Equity, a coalition of women's organizations, unions, other work force representatives, civil rights organizations and researchers, which coalesces on the issue of pay for work of comparable value. On the Federal level, all the major unions, Federally Employed Women, and others meet regularly to share information and develop programs. A similar vehicle for catalyzing work is appropriate here.

CONCLUSION

In sum, there is much that Federal sector unions are doing for clerical health and there is much more to be done. The demands for a healthier work life for office workers will continue to find a voice in unions. Just as slaves on the old plantation were considered happy by their owners, secretaries in their pink collar ghetto will be seen by their management as having it easy, until, through self-organization, they expose the injustices, reverse their invisibility, and organize for a better, more healthy quality of work life.

The film "9 to 5" captured the problems commonplace in clerical worklife and offered an image of an alternative. But the way to achieve flexitime, childcare, worker democracy is portrayed through a fanciful management genie who appears after the acting out of individualized revenge fantasies against an individual boss. The solution really lies in the nitty gritty work of collective action on the part of secretaries through unionization. This is true in the Federal work force as elsewhere.

REFERENCES

1 5 United States Code Section 1101
2 29 United States Code Section 151
3 29 United States Code Section 651
4 5 United States Code Section 7902
5 Executive Order No. 12196, 45 Fed. Reg. 12769 (Feb. 26, 1980)
6 29 Code of Federal Regulations 1969
7 29 Code of Federal Regulations 1960.1 - 1960.8
29 Code of Federal Regulations 1960.37 (a)(2)(iii'
Supra, Note 30 at p. 3
8 5 United States Code Section 7119

9 5 United States Code Section 7106
10 NIOSH Technical Assistance 80-39, September 1980, Brookhaven IRS Service Center
11 5 United States Code Section 7106 (b)(1)
12 R. A. Karasek, Jr., "Job demands, job decision latitude and mental strain: Implications for job redesign, Admin. Sci. Quarterly, pp. 285-308.

# SOLVING PROBLEMS TOGETHER: AN ALTERNATIVE TO SYMPTOMATIC HEALTH PROBLEMS IN SECRETARIAL AND CLERICAL JOBS

KRIS MacGAFFIN
Consultant
1650 Harvard Street, N. W., #309
Washington, D. C. 20009

## INTRODUCTION

In the recent literature concerning occupational stress, two facts have become apparent: that secretarial and clerical jobs, potentially among the best bread-and-butter jobs in the country, are today unnecessarily stressful; and, second, that secretarial and clerical employees know best where the unusual stresses are located. The on-going working group process for secretarial and clerical employees is an effective way of clearly identifying traditionally intractable problems among support staff and of finding practical, workable solutions.

## STRESSES OF SECRETARIAL WORK

Clearly, secretarial work is inherently stressful. The standard for the work, though rarely achieved, is perfection itself: perfect typing, proofing, and editing; perfect organization; perfect collecting and distributing of information, and so on. Place this occupational standard in its normal context, in a physical environment that is public, open to hundreds of interruptions and intrusions of great variety from minute to minute throughout the day, add the personal characteristics most employers want from secretarial staff--a certain cheerful ability to cope under all circumstances--and one has the formula for great stress. But this is _not_ the stress that is problematic. On the contrary, many secretaries thrive on just this kind of challenge. Many secretaries work under seemingly intolerable conditions-- poor machinery and materials; unattractive, crowded, windowless rooms; tense relationships with clients, and the like--yet are happy, hard-working, satisfied employees. Conversely, many secretarial staff whose working environment is the best, are suffering from a variety of symptoms of burnout and alienation.

The unnecessary stresses are harder to understand. These are the stresses that contribute substantially to the many health problems affecting secretarial and clerical employees today, ranging from physical illnesses--chronic back problems, eyestrain, and migraine headache--to physio-emotional problems--including alcoholism, drug abuse, lethargy, depression, moodiness, hostility, bouts of tearfulness and crying, high absenteeism for sickness and "family problems."

Many employers complain that when they ask their secretarial employees to describe their problems they are met with either silence or an emotional elaboration of "little" or "personal" problems that are not central to "the big problems" in the office, that are vague and too general for response, or that are simply unrealistic and have no solution. More than one well-intentioned supervisor has had the confusing experience of trying to encourage secretarial staff to speak their minds and get matters out on the table, with the result that support staff either become overly emotional or remain stubbornly silent.

When encouraged to vocalize their problems in a private and supportive setting, secretarial personnel generate lists of work-related issues about which they feel strongly. These issues may initially be personalized, vague, general, and highly emotional. Such a list may include the following grievances:

- they treat us like children
- they take us for granted
- they don't respect us
- they're sneaky and vindictive
- they treat us like machines
- they think we're their personal servants

When confronted by complaints of this sort, supervisors may become frustrated or even antagonistic and refuse to enter into constructive dialogue about the problems.

It is at this point that stress builds for secretarial and clerical employees. Denied the opportunity to talk out their problems and issues constructively and to be "heard" by managing staff, the secretarial staff begins to suffer from unnecessary stress that finds its outlet in either hostile, aggressive behavior, or in physical and physio-emotional health problems such as those described above.

## THE WORKING GROUP PROCESS

The communications gap, however, though formidable, can be bridged. For example, a lead secretary for the director of a large division in a Federal agency, was experiencing a variety of physical and physio-emotional problems, including daily headaches, depression, loss of self-confidence, chronic bouts of crying alternating with periods of incapacitating (but unexpressed) frustration and anger, and lethargy. An initially competent, diligent secretary who, she said, "basically loved secretarial work," her performance on the job was now deteriorating, as was her relationship with her boss. Her first response, when asked in a group session to vocalize her problems was, "My boss treats me like a child." Obviously under considerable stress, she could offer only this emotional statement, which told the group participants nothing about the specific nature of the problem.

Empathy and encouragement from the group, however, gradually resulted in her becoming more specific, and the following problem was uncovered. Six to eight months earlier, her boss had begun to suspect that some of his analytic staff were slouching on the job: arriving late, taking long breaks and lunches, using work time on personal research projects and other deals they had going outside the office. But he wasn't sure, and he was concerned that he not alienate his staff. Instead of working things out directly with his staff, he came up with the structural solution of appointing his secretary to be time-keeper for the office. Her efforts at this task, however, were not working out. Attempts to solicit cooperation from the analytic staff failed because they kept "forgetting" to tell her their whereabouts. Her attempts to guess at their comings and goings resulted in disputes. When she tried to get her boss to arbitrate differences, he either dropped the matter or supported the analysts' arguments. Still, her boss insisted that she continue to keep the time sheets.

Further discussion in the working group produced several positive results. First, she visibly relaxed merely as a result of being able to talk and of finding acceptance for her dilemma. Second, it became clear to her that she needed to find a way to return the problem to her boss's responsibility. Third, now that she felt more calm and self-confident, she also felt more _able_ to work things out with her boss.

There are many aspects to this case that are important here. First, the secretary was becoming ill and depressed as a result of feeling unable to straighten out her problem constructively. Second, her initial description of the problem _was_ vague, emotional, accusatory, and unworkable. Third, she was able, under the right circumstances, to become clear and specific about the specific problem. Fourth, her boss was attempting a structural solution to an interpersonal problem with his analytic staff, and it wasn't working out. Fifth, his secretary, having figured this out, knew she would be treading thin ice to try to pass it back up to him, especially feeling as though she had failed in her assignment, and so she internalized and disguised the whole issue, even for herself, by becoming "ill."

The hopeful information here is that seemingly intractable, vague problems expressing themselves symptomatically as physical and physio-emotional health problems in secretarial and clerical staff _can_ be clarified in the working group setting, and thus be solved. The unpleasant information is that the process requires encounters and confrontations that most secretaries and clerical staff as well as supervisory staff would much rather avoid.

The office is an ecology of unequals. This is a paradox. Everybody needs everybody else, but some people seem to have more power than others. Each position of power and powerlessness carries its own benefits and liabilities. Secretarial staff are always acutely aware of the managing staff's power to reward and punish. On the other hand, supervisory staff frequently express frustration and helplessness in the face of the secretarial staff's power to mysteriously control productivity, to foul up progress on work or access to individuals, or simply to quit when they are most needed.

There is a tendency, under these circumstances, to avoid solutions that require interpersonal risk, and to try to substitute impersonal, structural solutions. For example, a problem developed in the working relationship between a lawyer and his secretary; rather than address it directly, they collaborated in substituting communication by memorandum. They did not speak to each other for several months. However, this approach failed. The secretary, who could no longer endure the stress, nor the anticipation of a poor performance evaluation, was on the verge of quitting. This threat led the lawyer to seek help. They had both tried to avoid the unpleasant and frightening direct communication process to solving the problem and tried to substitute an impersonal, structural solution that did not address the causing problem. With the help of a consultant and much risk-taking on their part, they worked their way back to the original problem and sorted things out.

Impersonal, structural solutions include a wide range of measures, from the physical to the procedural, from simply moving "difficult" people around and relying heavily and somewhat punitively on rules, regulations, and policies, to solving problems by memoranda, petition, and special contract. The tendency, in an ecology of unequal power, seems natural: No one wants to "lose" the particular battle at hand. Most employees would rather line up as much impersonal power on their own side as possible and attempt to find sweeping, structural solutions that will do away with sticky problems once and for all. The problem with this is that many structural solutions tend to repress problems and difficulties rather than solving them; and they also tend to encourage an adversarial polarization of "sides" in an office. Once the struggle is polarized in a situation of unequal power, the party with less power tends to stubbornly and defensively dig in for survival, and then the whole problem is on an intractable, downward spiral.

It is very important to be sure that structural solutions address the problems at hand and solve them. Some Federal agencies are currently experimenting with carrying out routine quarterly performance appraisals instead of waiting out the year for the annual evaluation. Quarterly discussions with employees are certainly more informative than annual ones. It is a structural solution that arises out of a familiar problem: no one likes performance appraisals. Supervisors dislike doing them, and subordinates dislike getting them, especially if they are less than perfect. No one likes them because they are one of the rare responsibilities that actually require an interpersonal addressing of (potentially) unpleasant information. Indeed, the appraisal discussion must be required or most employees would simply avoid it altogether. And yet, ironically, the spirit of the appraisal system is positive: that employees and supervisors have an opportunity to discuss problems <u>before</u> they get out of hand. Still, the encounter is feared, and even with the quarterly appraisal program, many will find ways to make the experience as perfunctory and impersonal as possible.

This demonstrates that even structural solutions, if they are to be truly effective at <u>clearing up</u>, rather than avoiding the problem, always have an interpersonal aspect.

Interpersonal communication involving differences of opinion and idea in a relationship of unequals is extremely difficult for everyone, and yet we cannot avoid it if we want to clarify problems and find real solutions. For individual, isolated secretarial and clerical employees, who have the least real power in the ecological system of an office, the prospect of bringing up problems, differences of opinion and the like, and struggling with people in authority who have the power to fire, can be paralyzing. Indeed, it frequently seems suicidal even to whimper. One person alone behaving in a forthright way, unless she is extremely charming and self-confident, can appear to be a troublemaker and find herself out on the sidewalk. Who, after all, wants to be the first and only one out on the ice?

In one working group, the word processors began to complain openly in the group that they were frustrated by a problem in their office. It seems the electric current in their working space would not support the three word processing machines <u>and</u> the copying machine running at the same time. When someone was using the copying machine, the clerks had to turn off the word processing machines, sometimes resulting in the loss of the material on a disk. The management staff all had access to the copying machine and frequently used it. Each time, the clerks had to turn off their machines. None of them wanted to complain, for fear of being identified as having a "bad attitude," because, after all, the management staff was being nice by doing their own copying. None of the management staff was aware, however, of just how frequently the interruptions occurred, because the clerks were afraid to speak up. A tremendous amount of time was lost in this way, in addition to the stress the clerks experienced.

As a result of their participation in the working group, the clerks decided to monitor the amount of time their machines were "down" and the amount of time they lost putting down their work, getting distracted with other work, and getting back to the original task. With this homework done, they were then going to request a meeting with the appropriate management staff to present them with the problem and to offer a variety of solutions that can be discussed together.

Without the opportunity to talk together in their working group, to come to an agreement about their problem, to discover that in fact it might be appropriate (rather than troublemaking) to talk to the management staff about the problem, and to work together on developing a variety of solutions to the problem, this group of clerical employees would have remained frustrated, resentful, and stressed by their burden.

The working group model provides a regular, supportive forum for secretarial and clerical employees to discuss their problems and issues and to find encouragement, support, and assistance in addressing them as openly and positively as possible. The process is not easy, and the groups do not always succeed in establishing enough constructive rapport with each other so that problem-solving is possible. Certainly an experienced, supportive facilitator is essential to help the groups come together and to help them find a constructive working process that suits the group's needs. This working group model has been effective in improving working relationships and, therefore, offers benefits that justify further research and implementation in the work setting.

# SECTION 3
## Ergonomic Aspects of the Workplace

# ERGONOMIC ASPECTS OF HEALTH PROBLEMS IN VDT OPERATORS

MICHAEL J. SMITH
Division of Biomedical and Behavioral Science
National Insitute for Occupational Safety and Health
Cincinnati, Ohio

## INTRODUCTION

Concerns about video display terminals (VDTs) and their potential health implications originated in Europe during the 1970's. Some of the first research examining VDT operator health problems was performed in Austria by Haider and others (ref. 1, 2 ). They demonstrated that VDT operators who performed tasks under laboratory conditions displayed decreases in visual acuity in terms of increased myopia, which became more profound with increased viewing time. The greatest visual acuity changes reported were approximately 1/4 diopter. All operators recovered their previous visual function within 15 to 20 minutes. While the changes observed by Haider and others (ref. 1) were not very large, they demonstrated that short periods of viewing can produce changes in visual function. Because it is possible that these visual function changes may have long term effects given chronic exposures, recommendations were given on rest break allowances.

Laubli and others (ref. 3) in Switzerland showed that VDT employee visual complaints such as eyestrain carried over from one day to the next. This demonstrated a potential for chronic impact of VDT viewing on the visual system. In addition, they found that VDT operators performing different kinds of work activities displayed different levels of visual problems. Those operators performing record-retrieval tasks (interactive) had more visual complaints, such as eyestrain, than those VDT operators performing data entry tasks or typists using standard electric typewriters. Hunting and others (ref. 4) found that VDT operators' muscular complaints (sore shoulder, back pain, sore wrist) for these various jobs showed a different pattern from the visual complaints. For the muscular problems, the data entry VDT operators had the highest levels of complaints, with the interactive operators and standard typists reporting lower levels. Clearly, this demonstrates how job requirements can influence the effects of VDT use in defining the potential health problems of VDT work.

In Sweden, Gunnarsson and Ostberg (ref. 5) found that VDT operators at an airline reservation facility reported high levels of visual and muscular problems. For some of the visual problems, over 80% of the VDT operators reported a complaint. These findings led to the formulation of national health and safety regulations with provisions for VDT spectacles and rest break allowances. Ostberg and others (ref. 6) later showed that VDT operators' dark focus changed after VDT viewing, possibly indicating a shift in accommodation resting point.

A great deal of research dealing with ergonomic* aspects of VDTs has been carried out in Germany and has led to detailed national health and safety regulations on VDT design and work station requirements. Cakir and others (ref. 7) conducted a large scale study of over 1,000 VDT operators and a number of comparison occupations. A variety of factors ranging from health complaints to VDT design problems were examined. This study was the foundation of the document "The VDT Manual" authored by Cakir, Hart, and Stewart (ref. 8), which has been utilized worldwide by many practitioners for dealing with VDT problems. This study corroborated previous findings of increased visual and muscular problems for VDT operators. More importantly, it defined a number of design and environmental features that could have been responsible for the problems observed. Some examples of problems were screen glare, nonoptimal illumination levels, poorly designed furniture, and screen flicker.

The result of this work in Europe is that Sweden, Germany, and Austria have occupational health standards regulating VDT design and work activities. In addition, Norway will soon have standards enacted and Canada is holding hearings on national standards for regulating VDT use. In the U.S., a number of state legislatures are considering legislation to regulate VDT use.

## VDT HEALTH ISSUES

The National Institute for Occupational Safety and Health (NIOSH) first became actively involved in the possible health impact of VDTs in 1975 when ionizing radiation measurements were taken on terminals located at a newspaper on the east coast see (ref. 9). Since that time, NIOSH has undertaken a number of evaluations of radiation emissions from VDTs. For the VDTs examined, the emissions measured in the x-ray, ultraviolet, visible, infrared, and radiofrequency spectrums were less than current U.S. threshold limit values or standards, and thus believed to pose no hazard. Later, evaluations also made mention of potential ergonomic-job stress problems.

Employee complaints in VDT jobs concerned inadequacies in machine design, workload, rest breaks, environmental qualities, employee-supervisor interaction, work content, and social interaction.

---

* Ergonomics refers to the interplay between humans, machinery and the work environment and how each element influences the others during work activity.

One NIOSH field evaluation examined the potential radiation hazards and human factors problems for VDT operators at two newspaper and one insurance company in San Francisco (ref. 10, 11) This evaluation had three components. The first was a questionnaire survey to examine employees' health complaints. The second consisted of a field evaluation of basic ergonomic features of a sample of VDTs. Factors such as screen size, glare, lighting, postural requirements and machine/worker incompatibilities were examined. The third phase consisted of radiation measurements on a sample of VDts to identify the levels and sources of ionizing and nonionizing radiation.

The radiation testing measured no levels of radiation that exceeded current U.S. occupational radiation standards (ref. 11). The clerical VDT operators reported higher levels of visual complaints than either the professionals using VDTs or the nonoperators. These visual complaints included burning eyes, itching and tearing eyes, blurred vision and eye strain. Most of these complaints are symptoms of visual fatigue that could have been produced by prolonged viewing, glare from the VDT, and/or improper illumination levels. The clerical VDT operators also reported more muscular complaints than the professionals using the VDTs or the nonoperators. These complaints showed a generalized effect ranging from back pain to sore neck to wrist and finger soreness. Such complaints could be related to the keying requirements of the VDT work and possibly to poor posture that could be due to improper seating or incorrect visual viewing distance or angle. Both the visual complaints and muscular aches and pains are indicative of a generalized stress response related to machine and environmental design problems and job task demands.

Emotional difficulties were also noted. Both clerical VDT operators and professionals using VDTs reported higher levels of anxiety, depression, tension, and mental fatigue than the nonoperators. This emotional effect could be related to the job demands that required tight deadlines and resulted in pressure to produce. It is reasonable to assume that the three areas of health complaints (visual, muscular, and emotional) can be tied directly to the use of the VDT and to the job demands imposed by the work activity of the operators. However, the relative contribution of each of these factors is not known and needs to be identified, since each requires a different approach for remedial action.

The single major ergonomics complaint concerned screen glare, which was observed for a high proportion of VDTs at all of the work stations examined. Such glare could have caused the eyestrain reported by VDT operators. The glare was caused principally by poor placement of VDTs with respect to potential glare sources such as windows and lights and generally inadequatelighting design for VDT use. This was evident despite the fact that the levels of general illumination at all workplaces fell within recommended guidelines of the American National Standards Institute (ref. 12). This demonstrates that there are additional concerns about lighting in VDT work areas that are not taken into account by general office requirements, and added measures are essential to minimize glare. Other significant ergonomics problems related to VDT design and operation included: (1) no allowance for individual operator adjustment of such factors as screen brightness and contrast or screen and keyboard height, and (2) inadequate furniture for use with VDTs, such as nonadjustable chairs and tables. These factors could lead to improper work postures and produce muscular problems and visual fatigue.

In summarizing some of the research findings regarding VDT health issues, visual problems have predominated in the studies both in Europe and in the U.S. The evidence from these studies indicates that VDT operators, as a group, suffer from a high incidence of visual disturbances including visual fatigue, visual irritation, and headache. In addition, it is clear that the type of VDT work activity, such as data entry versus word processing, and the specific visual demands imposed by that activity, influence the incidence of visual complaints. VDT workers at visually demanding jobs (for example, records retrieval) have a much higher rate of visual complaints than those at less visually demanding jobs (for example, word processing). Furthermore, almost all types of VDT work activity produce higher levels of visual complaints than traditional office work that is also visually demanding. The same sort of pattern holds true for muscular problems. As with the visual complaints, they vary with the type of VDT work activity, but generally almost all types of VDT work produce more muscular complaints than other types of traditional office work. These muscular complaints are of a diverse nature affecting the neck, the shoulders, the back, the arms, the hands, and the fingers, which demonstrates an effect on the total musculature. It must be pointed out that the nature of VDT work is typically so different from traditional office work that it is futile to try to make comparisons in terms of physical job demands, and this comparison to standard typing activities is not meaningful.

Another area of health complaints reported by the VDT operators concern emotional disturbances. These fall into two general categories: those that reflect mood disturbances, and those that reflect psychosomatic symptoms. Very few studies done on VDT operators have examined these issues. The NIOSH San Francisco study (ref. 13) and work by Elias and others (ref. 14) in France both indicate mood disturbances and psychosomatic symptoms for VDT operators. These symptoms again vary by the type of VDT work activity, and as with other types of health complaints, VDT operators have more of these types of symptoms than other types of office workers. The mood disturbances typically are of a neurotic nature including anger, frustration, irritability, anxiety, and depression. The psychosomatic disorders reflect a typical distress syndrome, including gastrointestinal disturbances, muscle and psychic tension, heart palpitations, and frequent sweating.

The final area of VDT health concerns deals with psychosocial disturbances. VDT operators in a number of the studies complained about specific job demands such as workload, work pace, and supervision style, which then manifested themselves in health complaints of both a physical and psychological nature, as well as general job dissatisfaction. As with other types of health complaints, the psychosocial problems also vary with the type of VDT work activity. These complaints are of particular importance since they have a relationship to distress syndromes and thus coronary heart disease, gastrointestinal problems, and psychological disorders.

Factors contributing to observed health complaints

Since VDT operators have a high incidence of some types of health complaints, what factors have been identified in studies as contributors to these problems?

(i) Environmental design. Three of the most often cited causes of vision problems are improper illumination, glare on the VDT screen, and improper contrasts in luminance. These environmental problems usually occur together, although any one could conceivably produce visual disturbances. Hultgren and Knave (ref. 15) found lighting levels at 76% of the work stations examined exceeded 700 lux (30% over 1,000 lux), while at 53% of the work stations VDT operators reported disturbing reflections on the screen and 59% of the VDT operators reported trouble reading images on their VDT screens. Findings similar to these have been shown by Cakir and others (ref. 7) and Laubli and others (ref. 3) However, variations have been observed in some studies. For instance, Gunnarsson and Ostberg (ref. 5) found that illumination levels varied between 150 and 500 lux. In the NIOSH San Francisco study (ref. 16) over 85% of the work stations examined had illumination levels between 300 and 700 lux which are considered to be appropriate illumination, depending on the VDT task. As with the other studies, bothersome glare was observed on their work station, 85% of the VDT operators cited screen glare, 70% character brightness, 69% readability of screen, 68% screen flicker and 62% screen brightness. Coe and others (ref. 17) also found that 90% of the work stations examined had lower illumination levels than those observed in European studies (500 lux or lower). In fact, it was observed that there were more visual complaints when the lighting level was below 500 lux than when it was above 500 lux.

However, again glare was a significant problem (42% of the VDTs). It seems that VDT operators still complain about visual problems when illumination levels are below 500 lux and above 200 lux. Thus, it would appear that glare and/or contrast problems are more critical elements in VDT operator visual complaints than illumination level. This was confirmed by Laubli and others (ref. 3) who found a correlation between the measured intensity of glare reflections and reported annoyance by the VDT operators, but no relationship between the luminance of the reflections and reported visual impairment. On the other hand, Stammerjohn and others (ref. 16) found a clear relationship between glare and visual complaints.

Temperature and humidity are two areas that always seem to evoke complaints from office workers. Dissatisfaction with these workplace features may influence worker perception of the overall job situation. Likewise, it is not known whether office atmospheric conditions and/or static electric field created by VDTs have direct influences on worker health. Cakir and others (ref. 8) found, that in non-air conditioned offices about 50% of the VDT operators complained of the heat while in air conditioned offices, 30% complained of the heat. In addition, almost two-thirds of the VDT operators complained that the air was too dry, even though the relative humidity in their workplace was between 30 and 40%. In the NIOSH San Francisco study (ref. 16), 63% of all employees rated summer temperatures as too high, while 41% rated winter temperatures as too low, even though measurements of the temperatures and relative humidities at the worksites were within established limits for comfort (between 21 and $25^{o}$ C and 35 and 80% relative humidity). Coe and others (ref 17) found that 80% of the VDT operators and 75% of the controls reported that the temperature of their work areas was uncomfortable.

(ii) Work station design. A host of features of the VDT, the desk, and the chair have been examined to determine relationships to health complaints. In terms of the screen, the issue of adequate contrast between the characters and screen background is of concern. In many of the field studies, these contrast ratios were much less than optimal (7:1 to 10:1 is considered acceptable), and in some cases were less than adequate (3:1 is considered adequate) (ref. 16). Grandjean (ref. 18) has demonstrated that the quality of the screen characters can influence the level of health complaints. Although, most field studies have found that the large majority of VDTs examined have the capabilities to meet minimum requirements for contrast ratios and character size, many do not achieve these minimum requirements in the field as a result of screen glare.

The work station has been implicated in a number of the studies as a significant factor contributing to both visual and muscular health complaints.

In particular, the height of the working surface has an impact on the height of the arms, wrists, and hands, as well as wrist and neck angles (ref. 19). In addition, the height of the chair and the amount of support that it provides for the lumbar region of the back have been postulated as factors contributing to worker muscular complaints. Studies by Stammerjohn and others (ref. 16), Cakir and others (ref. 7), Coe and others (ref. 17), and Sauter and others (ref. 20), all demonstrate that a majority of VDT operators are exposed to undue muscular loads (postural and manipulative) because of poorly designed, or improperly used, work station furniture.

(iii) Job design. If all ergonomic aspects of the environment and work station were maximized, we would still find VDT operators with visual and muscular complaints. Job demands, both physical and psychological, influence the type, severity, and frequency of VDT operator health complaints. It is also logical to assume that psychological demands can produce physical as well as emotional complaints. Some support for this view is suggested by the fact that, in three major studies (ref. 5, 10, 20), the amount of time working at the VDT was not very predictive of the frequency of visual or muscular health complaints, while the types of VDT work activity was, with the more mundane, boring, repetitive VDT jobs, showing higher levels of both visual and muscular problems.

The evidence as to whether VDT operators have greater job demands than other occupations is mixed. In most cases, the evidence is confounded by other ergonomic factors, as well as the lack of adequate study designs. Gunnarsson and Ostberg (ref. 15) found that when VDT operators had little control over job tasks, 72% complained of monotony. In another group, with greater control, only 10% complained of monotony. NIOSH (ref. 10) evaluations indicated that clerical VDT operators reported less work involvement, job autonomy, self-esteem, staff support, and peer cohesion, with greater work pressure, workload, workload dissatisfaction, and more supervisory control, role ambiguity, and job future ambiguity than professional VDT operators. For all these dimensions except staff support, job involvement, and role ambiguity, the clerical VDT operators also indicated more stressful responses than clerical control subjects. Coe and others (ref. 17) found differences in the level of work pressure between VDT operator groups in the newspaper industry. Editors showed the most work pressure and interactive operators showed the least. However, control subjects showed greater work pressure than all VDT operators, except the editors. For boredom and frustration, there were no differences between the groups.

Sauter and others (ref. 20) found that a heterogeneous group of VDT operators did not differ greatly from clerical controls on similar dimensions of stress than those reported by NIOSH ref. 10). However, the job stress factors were highly predictive of health complaints, job satisfaction and emotional status for both VDT operators and clerical controls.

It appears that job design factors, primarily highly paced work, lack of control, and pressures such as deadlines are more prevalent in some types of VDT work than in others. These factors have been shown to be related to many of the health complaints that are reported by VDT operators and could be linked to health disorders.

(iv) Organizational design. The organizational factors that have been associated with VDT work as sources of distress are: (1) lack of worker participation in VDT implementation, (2) inadequate employee training, (3) job security issues (such as downgrading, advancement and job loss), (4) the monitoring of employee performance, (5) the influence of close employee monitoring on supervisory style, and (6) incentive pay schemes. These have been identified in studies by Cakir and others (ref. 7), Smith and others (ref. 10), Johansson and Aronsson (ref. 21) and Sauter and others (ref. 20).

## Ergonomic solutions to VDT problems

Given the research findings on VDT problems, it is possible to establish a framework from which appropriate ergonomic interventions for various types of VDT work activity can be examined. First of all, there are four basic points when considering ergonomic issues that relate to the basic functions of VDT use, and thus the VDT work system. The first deals with the work environment. Where is the VDT located? Where is it being utilized? What are the conditions of the work environment? The second deals with the extent of use. How long is the VDT used: 4 hours a day? 8 hours a day? The third deals with the types of tasks. How are the tasks done in terms of the thought requirements versus the perceptual motor requirements? And fourth, what are the psychological loads imposed on the individual in carrying out this work? Is the work boring? Does the work present a challenge? Does the work have meaning to the individual?

These points provide the basis for looking at four areas of health issues in relation to VDT work. These areas include the visual load, the muscular load, the postural load, and the emotional stress load. Visual load is reflected in visual fatigue, eye irritation, and possibly headache. Muscular load is reflected in muscular fatigue, sore muscles, and sore joints of the munipulative components of the body, such as the arms, the hands, the wrists, the fingers, and the neck. Postural load, on the other hand, is reflected in muscular fatigue and sore muscles and joints related to the major body musculature that keeps one seated upright, such as the back muscles and the leg muscles. The emotional load is reflected in emotional distress such as anxiety, depression, irritability, and fatigue, and in such reactions as job dissatisfaction. This framework for examining the various load aspects of VDT work and the strains they impose defines how interventions for the different categories of VDT work can be examined.

(i) Data entry tasks. Data entry jobs are typically hard-copy oriented, and thus the operator spends little time looking at the screen. These jobs are characterized by high rates of keying with very little time viewing the VDT. An example of the typical data-entry operation would be claims clerks at an insurance company. Many such jobs limit the clerk to only entering information about the insurance claims into the computer system. The clerk does not ordinarily have to look at the screen, but just works from hard copy, keying the claims information and looking at the screen only if he or she thinks a mistake has been made. This type of job is like old-fashioned keypunching; however, there is the addition of the video display screen which tends to attract the eyes away from the hard copy.

In examining three of the categories of impact, that is visual, muscular, and postural loads, for data-entry jobs, it is obvious that the visual load is not related to the screen, since this is not a screen intensive job. In this case, the visual load is related to the hard copy, which can create as much visual fatigue as the VDT screen itself because the quality of the hard copy is often poor. Since the hard copy is the major factor that relates to the visual issues, it is the factor that must be addressed. If we use lighting recommendations that have been proposed, this type of job would require 50 foot candles (approximately 500 lux) of illumination (ref. 22).

In terms of the musculature, these jobs impose a very great muscular load. In particular, there is a heavy load on the hands, wrists, fingers, and arms owing to the number of keystrokes required per hour. An ergonomic solution for dealing with this situation is to take the load off the fingers and the wrists by using a wrist rest. Additionally, it is important that the position of the wrists and the arms for these individuals be correct, and thus adjustable keyboard height and position are very important. To achieve this, an adaptable work station is necessary.

When examining the postural load, it is observed that there is a great load on the back and shoulders for operators who are doing very intensive keying. For this situation, the most important feature is a proper chair. The chair must provide adequate lumbar support. The chair height should be adjustable, as well as the back rest tension. In addition, this job requires handling and looking at hard copy, which produces a good deal of postural adjustments and head and neck movements. In order to reduce the number of neck problems, it is important to have a document holder to reduce the number, pattern, and extent of such neck movements. To summarize, for data entry types of jobs, the screen is secondary, while the work station adjustability is primary. Therefore, it is important to provide a good chair, an adjustable keyboard, and a proper surface for source documents.

(ii) Data-acquisition tasks. Data acquisition tasks are almost the opposite of data-entry tasks in terms of the load characteristics. The kinds of jobs in this category are telephone directory operators, air traffic controllers, and postal clerks working on the computer mail forwarding system. These are jobs where operators spend most of their time looking at the screen to retrieve information. It can be a visually demanding, screen-intensive task, requiring the best available VDT screen. Thus, good character quality, high character-to-screen contrast, and glare control are essential (ref. 23). In addition, environmental factors are important such as lighting, environmental glare control, and the viewing distance. In terms of illumination, it has been suggested that this work activity requires about 30 foot candles (approximately 300 lux) of illumination (ref. 22).

In terms of muscular load, it is difficult to offer a general specification for data-acquisition VDT operators. Telephone directory operators appear to have high muscular and postural loads because of repetitive work at a very quick rate. On the other hand, air traffic controllers have a completely different type of situation, with a high cognitive and perceptual load but with a lesser muscular load. In those job tasks where the load is high as a result of a high number of repetitive motions, there is a need for a wrist rest and an adjustable keyboard. In terms of postural load, almost all retrieval VDT jobs have a high postural demand and thus require a good chair as well as adjustable screen height.

(iii) Interactive tasks. Interactive type of VDT operators have an astounding variety of VDT jobs, with a large number of different job tasks. This varies from computer programmers and Computer Aided Design (CAD) engineers, who work at very creative activities and may be able to get up and move around at will, to reservation clerks at airlines, who sit for hours continuously. Thus, there is a difference in terms of demands that are put on various types of interactive operators in terms of the viewing and the muscular requirements.

For a programmer, the visual load makes the greatest demand, while the muscular load and the postural load are secondary. In terms of reservation clerks, there is a dual load, visual load because of the need to be able to interact with the screen, and a muscular load because of the demand to be able to process information in a very rapid way with many keystrokes. Both, however, have a relatively high postural load.

For a highly visually demanding job, there is a need for good screen characteristics as discussed earlier. In addition, there is the need for a good visual environment with approximately 30 to 50 footcandles (300 to 500 lux) of illumination and good glare control. For most interactive jobs the muscular load is not as intense as it is for operators who do nothing but data entry. However, this load may still be relatively high and may make a wrist rest vital. In terms of postural load, an adjustable chair is required. Depending on the need to interact with the screen and the amount of viewing time, an adjustable screen height may be necessary. For particular applications, there is also a need for a document holder. This is more important for people who are using resource materials. Thus, the people in computer aided design may have a high muscular and postural load when using a light pen, but would not require a document holder. On the other hand, those working as programmers most likely would require a document holder of the appropriate dimensions.

(iv) Word processing. Here again, the loads will vary depending on the tasks that are being conducted. The kinds of jobs that are included here vary from typists, who essentially spend most of their time keying with very little time looking at the screen, except for corrections; to editors, who spend most of their time looking at the screen and very little time keying. Thus, we have a situation similar to the interactive jobs. The point is that the job task characteristics define what the ergonomic intervention should be and that major efforts need to be directed to where the major loads are going to be. Some jobs have dual loads on both vision and the muscular/skeletal system; some jobs do not have dual loads, and only affect the visual system or only the muscular system. In essence, the loads should define how we go about intervening in a work situation.

## Job design solutions to VDT problems

It is clear from the literature that the greatest job design difficulties in VDT work occur for jobs that inherently had little job content before automation; and even less with automation, such as insurance claims processing and telephone directory work. For work activity to provide adequate job satisfaction, there must be meaningful content for the individual to derive a sense of accomplishment and a positive feeling of self esteem. Many clerical VDT jobs are fragmented and simplified versions of traditional clerical activites. Thus, the little content that was in these clerical jobs has been diminished further by computerization, such that very little meaning or satisfaction can be derived from conducting the work activities. Therefore, boredom and fatigue predominate in these jobs. To enhance this type of activity, meaning has to be built into the job content. This can be accomplished by enlarging the use of workers' skills as opposed to simplifying the work. Work should not be overly repetitive to the extent that the VDT operator uses only simple perceptual motor skills and no social or cognitive skills. In addition, job tasks should be designed to utilize existing skills as much as possible to enhance worker confidence and performance.

Control of the work process is a significant factor in the occurrence of job stress. Lack of job control has been demonstrated to be one of the primary causes of psychological and physiological dysfunction (ref. 24, 25, 26) and looms as one of the major characteristics of computerized work processes. Providing a greater amount of control over their own work activities by increasing operators' decision making and use of alternative work procedures, reduces the stress imposed by computerized work processes. It also enhances the job content aspects of the work activity by giving more individual meaning and satisfaction with individual accomplishments.

Feedback about performance is a significant aspect of worker control of the work process. If the computer is continuously reporting performance information to the supervisor, who then uses this information to intimidate the employee, then the employee will perceive a lack of control over the work process. Rather than providing performance feedback to the supervisor, it may be better to give this information directly to each employee on his own VDT screen on a frequent basis. There is a large body of literature that indicates that such direct performance feedback to the employee has a positive influence on performance (ref. 27). On the other hand, having the supervisor provide the performance feedback may create tension and stress, and thus have a negative influence on performance.

Completeness is another aspect of job content and the meaningfulness of work that is often missing in computerized office work. As work is fragmented through simplification, the relationship of the task activity to the organization and end product is diminished. Thus, workers fail to identify with the work process and product. They fail to appreciate that a lack of quality in their small component of the product can have a major impact on the completed product and their fellow co-workers' performance. Fragmentation of work must be avoided, so that employees can attain a personal identity with the organization, a product identity, and an organizational pride. If work tasks absolutely have to be simplified and fragmented, then it is imperative that employees understand their contribution to the end product and the organization. They must feel that their contribution is significant and meaningful for a positive feeling of self esteem. Otherwise, health and productivity will suffer as a result of increased job stress. The operators' tasks should be broad enough to provide some closure and therefore understanding of the significance of the work.

Computerized systems often produce an isolation of individual operators to a much greater exent than traditional nonautomated processes. This isolation at a fixed work station greatly reduces social interaction, which has traditionally been one of the major benefits of clerical office work. As this social support is removed by isolation, the positive benefits in stress control are eliminated. However, with the use of VDTs, it is often not possible to have social interaction during the work task activity. Therefore, social interaction during nontask periods must be enhanced and encouraged. This can be accomplished by providing special work break facilities in close proximity to the work areas and by allowing groups of workers to go on break together.

The final job design issue deals with the determination of reasonable workload for VDT operations. The research literature has shown that workload for VDT operators is often set by the limits or capacity of the computer system, rather than by the capacity of the operator (ref. 10). This occurs because computer operations are set up by systems analysts and computer programmers, who have an understanding of the computer system capabilities but no idea of what people are capable of doing. Given that computer systems are an expensive investment for any company, it is understandable that production goals are often set based on the cost of the computer system and the need to improve productivity. However understandable this action, it is not based on sound engineering or psychological principles for determining the proper workload and thus often produces excessive workload. To determine the appropriate workload, industrial psychologists and industrial engineers, as well as employee representatives, must be part of the team that designs and implements the computer system.

## Organizational solutions to VDT problems

A major factor that produces worker resistance to automation is that automation often appears at the workplace "out of the blue" without worker knowledge of the impending change in the work process. It is very important for the successful implementation of computer automation and subsequent enhancement of worker health, performance, and satisfaction, that organizations have a transition policy that includes worker participation in all stages of the automation process. That is, workers should participate from the first decisions about whether to automate, through selection of equipment, to the daily operation of the computer system.

First, worker representatives should be involved in the planning phase of automation. This will aid in employee acceptance of the changes in work processes and ensure that employee concerns are aired. Secondly, the employee representatives should be involved in the design of the computer system, to ensure that human concerns and capabilities are included, as well as the computer capabilities. Finally, employees who are affected by the automation should be involved in the implementation of the automation. This will provide them with a fuller understanding of the computer system, its capabilities, and their role in the work process.

Training of the operators in their new job requirements is one of the most neglected aspects of computer automation. Of course, all companies provide an introduction to operators about how the equipment operates and what the various features are. Additionally, the manufacturer typically provides a manual that explains how the system works, and how specific functions can be carried out. However, the extent of operator training often is to tell the operator to read the manual and to start working.

Because VDT technology represents a completely new way of carrying out the work activity, the need to have comprehensive training procedures that will develop skills and enhance worker confidence and self-esteem is increased. Training should start with an explanation of why the new technology is needed, its benefits to the company, and its benefits to the worker. The equipment and computer system should be thoroughly explained, indicating the strengths <u>and</u> weaknesses of the system. Then, there should be intensive training from the manual that explains how the system works and specific functions. Each classroom teaching or individual reading session should be followed up by practice in the functions learned, with a skilled operator available to coach the trainees.

After the trainee has successfully passed the training course, he or she should not be required to work at full speed until becoming accustomed to the work situation. This could take from one day to a month, depending on the complexity of the work activities. In addition, all operators should have periodic retraining (at least every 6 months) to keep skills and confidence at peak levels.

While this training regimen is more complex than usually undertaken, it will enhance operator skills and will reduce the psychological fears of obsolescence and job loss, since trained operators will perceive themselves as important investments.

Fear for job security is one of the greatest concerns of VDT operators. This is natural since it is commonly believed that automation displaces workers. As indicated above, a company that invests time and energy in developing the skills of workers shows a desire to keep a valuable resource and thus reduces worker fears of job loss. However, there are other job security problems related to computerization. One is the possibility of being downgraded because the computer system takes over some of the worker's functions, making the job less complex. There is always a temptation to reduce labor costs by simplifying work. However, such efforts are almost always doomed to failure, because they produce extensive morale and motivation problems in workers, that directly influence productivity. The main purpose of computerization is to do work more efficiently and productively, not just to simplify it. Thus, companies should establish a policy that workers not be downgraded when VDTs are installed, since the workers will be more productive than they were before computerization. Ideally, the VDT should be the operator's tool rather than being a machine served by the operator.

In addition, companies must develop career paths for VDT operators so that advancement can be attained for those who are good performers. Being locked into a highly repetitive job, that has very little content or meaning with no chance for advancement is a major source of job stress and a demotivating force for many VDT operators. Companies will have to be innovative in advancing operators through a career path to enhance performance and reduce job stress.

It is quite clear from the literature that employee monitoring by computers creates a dehumanizing work environment in which the worker feels controlled by the machine. When performance monitoring is then used by supervision to control performance, worker perception of work pressure and workload is very high, thus producing stress responses. It also creates an adversary relationship between the supervisor and the employee. This is especially troublesome, since the supervisory/employee relationship may have been quite positive before the introduction of computerization and performance feedback.

For the most effective employee performance and to enhance stress reduction, supervisors should use positive motivational and employee-support approaches. This suggests that first-line supervisors should not be involved directly in the performance feedback system. Secondly, supervisors should be skilled operators who can assist those operators having technical difficulties. Thirdly, supervisors should receive training in employee support approaches, thus helping to buffer the effects of other stressful job demands. Companies must understand that this approach changes the basic role of the first-line supervisor by removing production pressures from the supervisor and establishing this individual as a positive link between employees and other levels of management.

## SUMMARY

There is a body of evidence that indicates working with VDTs can produce a variety of health complaints including visual fatigue and discomfort, muscular aches and pains, emotional disturbances, psychosomatic symptoms, and job dissatisfaction. A number of factors have been identified as contributing to these problems, such as improper lighting, environmental glare, poor work station design, inadequate chairs, poor job design, unreasonable organizational demands, poor supervision, and inadequate training. All of these influences are modified by the type of VDT activity being undertaken and the amount of load each puts on the visual, muscular, and emotional mechanisms of the operator. However, all can be controlled through the application of ergonomic principles for environmental, work station, job, and organizational design.

Computerized technology has the capability to greatly enhance the jobs of office workers by reducing undesirable, repetitive work tasks that require little thought and increasing the content of jobs by providing greater task variety and meaning. However, this technology also has the opposite capability; that is, to reduce office jobs to assembly-line systems in which job meaning is lost. The overall impact of the decision to enhance or degrade jobs through computerization will influence the economic benefits of computerization, the social and cultural development of segments of society and the health of the affected workers. In order to protect the health of those office workers affected by computerized technology, a management philosophy has to be adopted that will encompass considerations for (1) adequate environmental and work station design, (2) elimination of stress-producing job demands, and (3) provisions for more meaningful and positive work experience with needed worker training and career development.

## REFERENCES

1. M. Haider, J. Hollar, M. Kundi, H. Schmid, A. Thaler and N. Winter, Stress and strain on the eyes produced by work with display screens: report on a work-physiological study performed for the Union of Employees in the Private Sector. Austrian Trade Union Association, Vienna, 1975.
2. M. Haider, M. Kundi and M. Weissenbock, Worker strain related to VDUs with differently colored characters, in E. Grandjean and E. Vigliani (Eds.), Ergonomic Aspects of Visual Display Terminals, Taylor and Francis Ltd., London, 1980, pp. 53-64.
3. T. Laubli, W. Hunting and E. Grandjean, Visual impairments related to environmental conditions in VDU operators, in E. Grandjean and E. Vigliani (Eds.), Ergonomics Aspects of Visual Display Terminals, Taylor and Francis Ltd., London, 1980, pp. 85-94.
4. W. Hunting, T. Laubli and E. Grandjean, Constrained postures and VDU operators, in E. Grandjean and E. Vigliani (Eds.), Egonomics Aspects of Visual Display Terminals, Taylor and Francis Ltd., London, 1980, pp. 175-184.
5. E. Gunnarsson and O. Ostberg, The physical and psychological working environment in a terminal-based computer storage and retrieval system, National Board of Occupational Safety and Health Report 35, Stockholm, 1977.
6. O. Ostberg, J. Powell and A. C. Blomkvist, Laser optometry in assessment of visual fatigue, Department of Human Work Sciences, University of Lulea Report 1980: 1 T, Lulea, Sweden, 1980.

7. A. Cakir, H. Reuter, L. VonSchmude and A. Armbruster, Unterschung zur Anpassung von Bildschirmarbeitsplatseh an die Psysche und Psychische Function. Bundesministerinm fur Arbeit und Sozialordnung, 1978.
8. A. Cakir, D.J. Hart and T.F.M. Stewart, The VDT Manual. Ince-Fiej Research Association, Darmstadt Federal Republic of Germany, 1979.
9. C.E. Moss, W. E. Murray, W. H. Parr, J. Messite and G. J. Karches, An electromagnetic radiation survey of selected video display terminals, National Institute for Occupational Safety and Health Publication No. 78-129, Cincinnati, Ohio, 1978.
10. M.J. Smith, B.G.F. Cohen, L. Stammerjohn and A. Happ, An investigation of health complaints and job stress in video display operations, Human Factors, 23, (1981), pp. 387-400.
11. W.E. Murray, C.E. Moss, W.H. Parr and C. Cox, A radiation and industrial hygiene survey of video display terminal operations, Human Factors, 23, (1981) pp. 413-420.
12. ANSI, American National standard practice for office lighting: ANSI A132.1. New York: American National Standards Institute, 1973.
13. M.J. Smith, L.W. Stammerjohn, B.G.F. Cohen and N. Lalich, Job stress in Video display operations. in E. Grandjean and E. Vigliani (Eds.), Ergonomic Aspects of Visual Display Terminals, Taylor and Francis, Ltd., London, 1980, pp. 201-210.
14. R. Elias, F. Cail, M. Tisserand and M. Christman, Investigations in operators working with CRT display; relationships between task content and psychophysiological alterations, in E. Grandjean and E. Vigliani (Eds.), Ergonomics Aspects of Visual Display Terminals, Taylor and Francis, Ltd., London, 1980, pp. 211-218.
15. G. Hultgren and B. Knave, Contrast glare and reflection disturbances in the office environment with display terminals, Arbete och Halsa, (1973).
16. L. Stammerjohn, M.J. Smith and B.G.F. Cohen, Evaluation of work station design factors in VDT operations. Human Factors, 23,(1981), pp. 401-412.
17. J.B. Coe, K. Cuttle, W.C. McClellon, N.J. Warden and P.J. Turner, Visual display units, New Zealand Department of Health Report No. W/1/80, Wellington, New Zealand, 1980.
18. E. Grandjean, Ergonomics related to the VDT workstation, in Proceedings of Zurich Seminar on Digital Communication, Federal Institute of Technology, Zurich, 1982.
19. S.L. Sauter, G.E. Harding, M.S. Gottlieb and J.J. Quackenboss, VDT computer automation of work practices as a stressor in information-processing Jobs: Some methodological considerations, in G. Salvendy and M.J. Smith (Eds.), Machine Pacing and Occupational Stress, Taylor and Francis, Ltd., London, 1981, pp. 355-360.
20. S.L. Sauter, M.S. Gottlieb and K.C. Jones, A general systems analysis of stress-strain in VDT operations, Paper presented at the Conference on Human Factors in Computer Systems, Gaithersburg, Maryland, 1982.
21. G. Johansson and G. Aronsson Stress Reactions in Computerized Administrative Work, University of Stockholm, Stockholm, 1980.
22. W. Cushman, Illumination considerations for VDT work, Proceedings of Symposium on Health and Ergonomic Considerations of Visual Display Units. American Industrial Hygiene Association, Akron, Ohio, 1983, 73-86.
23. H. Snyder, Optimizing the equipment and operator. Proceedings of the Symposium on Health and Ergonomic Considerations for Visual Display Units, American Industrial Hygiene Association, Akron, Ohio, 1983, 21-28.
24. M.J. Smith, Occupational stress: An overview of psychosocial factors. in G. Salvendy and M.J. Smith (Eds.), Machine Pacing and Occupational Stress. Taylor and Francis, Ltd., London, 1981, pp. 13-19.
25. R. Karesek, Job demands, job decision latitude, and mental strain: Implications for job redesign, Administrative Science Quarterly, 24, (1979) pp. 285-308.

26. C.L. Cooper and J. Marshall, Occupational sources of stress: a review of the literature relating to coronary heart disease and mental illhealth. Journal of Occupational Psychology, 49, (1976) pp. 11-28.
27. K.U. Smith and M.F. Smith, Cybernetic Principles of Learning and Educational Design. Holt, Rinehart and Winston, Inc., New York, 1966.

# RESEARCH ISSUES IN THE ERGONOMICS, BEHAVIORAL, ORGANIZATIONAL AND MANAGEMENT ASPECTS OF OFFICE AUTOMATION

Gavriel Salvendy

Human Factors Program
School of Industrial Engineering
Purdue University
West Lafayette, Indiana 47907

## 1. WHAT IS OFFICE AUTOMATION?

The McGraw-Hill Dictionary of Scientific and Technical Terms defines office automation as "use of an electronic computer or computing system for routine clerical jobs". Such a definition is, of course, far too narrow to define the true scope of office automation of the 80's. Webster's Third New International Directory of the English Language (unabridged) provides a much broader and more representative definition for office and automation. Automation is defined as "The technique of making an apparatus (as a calculating machine), a process (as manufacturing), or a system (as of bookkeeping) operate automatically ... automatically controlled operations, of an apparatus, process, or system by mechanical or electronic devices that take the place of human organs of observation, effort and decision".

## 2. COMPUTER TECHNOLOGY AND ITS IMPACT ON OFFICE AUTOMATION*

The information processing industry as we know it today, with revenues exceeding $100 U.S. billion and affecting virtually every part of our national life, had its origin in a few powerful ideas. Among them were the binary number system, Boolean algebra, the mechanical adder, the punched card, the Jacquard loom, the Hollerith tabulator, the stored program, and the concept of a high-

* This material was initially presented as an invited plenary address at the 8th International Ergonomics Congress and subsequently published in Ergonomics (ref.13).

level programming language. Complex technological and social forces accelerated the development of these ideas and led to the modern computer, but the most influential force by far was economic demand. The computer made it possible to do the same work for less money. As the cost decreased and the function increased, use of the computer was rapidly extended to new and different kinds of work (ref. 9).

Figure 1 illustrates that in the past 30 years the price per one second of instruction has decreased drastically such that a computation which now costs \$1 would have cost about \$30,000 in 1950. If the automobile industry had incorporated equivalent cost reductions, one could now buy a new car for less than \$1. This marked reduction in the cost of calculations was closely associated with significant increases in memory product density (Figure 2). In effect, during the past 15 years the memory product density has increased such that an area which in 1965 could accommodate only one bit, now will accommodate over 1000 bits. These technological innovations of increases in density of bits per unit area and decrease in cost of computations were largely responsible for the emergence and rapid growth of the micro-computers for business and personal use. The fact that circuit reliability has also increased 1000 fold (Figure 3) has contributed to the wider acceptance and use of computer technology such that in 1980 \$355 per capita was spent in the United State on data processing which accounted for 5.2% of the gross national product. It is predicted that by the year 1995 these figures will be \$2,400 per capita spending and 21% of the GNP (Figure 4).

## 3. TECHNOLOGY OF OFFICE AUTOMATION

The technology of office automation has evolved over a number of centuries and some of the discoveries which have contributed to this process are listed in Table 1.

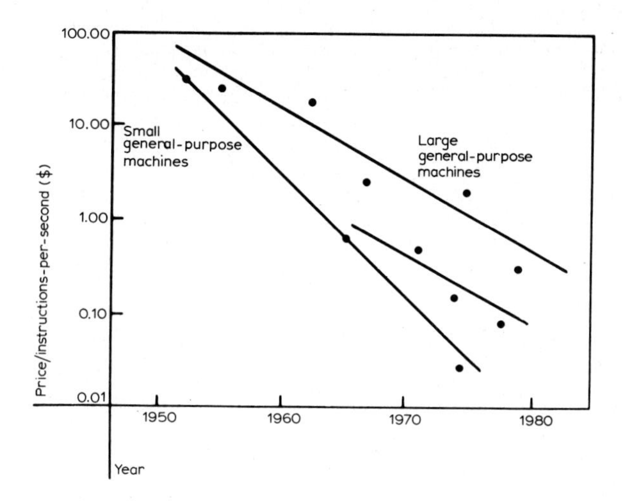


Fig. 1. Computer price-performance trends. Curve 1 represents an average improvement for large general-purpose machines of 15% per year; Curve 2 represents an average improvement for small general-purpose machines of 25% per year. (After Olsen and Orrange 1981.)

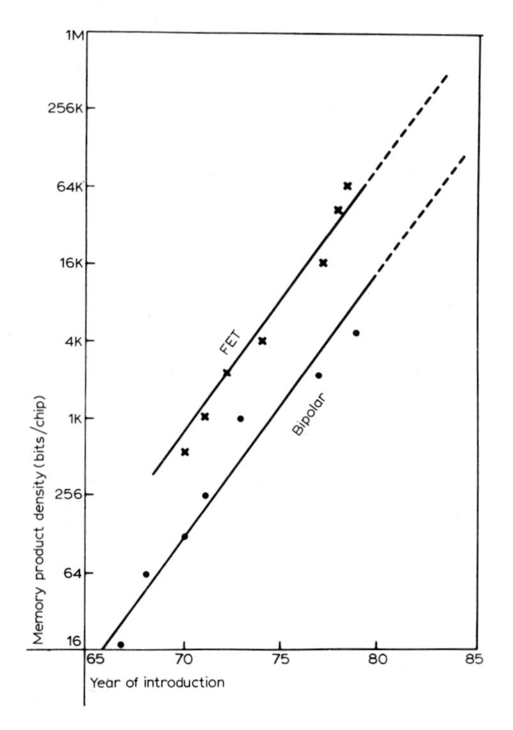


Fig. 2. Productivity growth for bipolar and FET memory chips. The dots and crosses mark actual products. (After Harding 1981.)

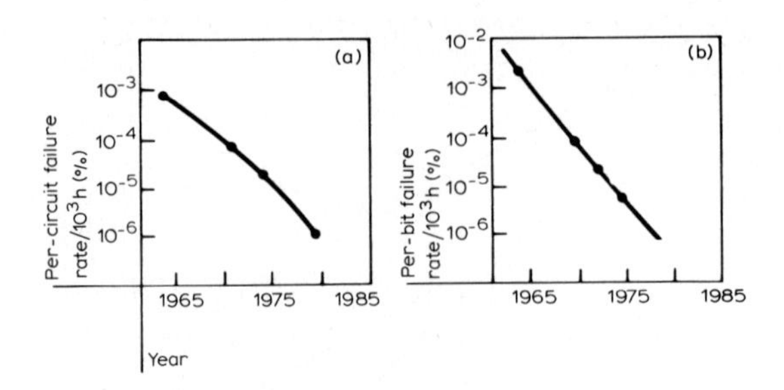


Fig. 3. Intrinsic failure rate improvement trends for IBM technology. (a) Logic circuit reliability, where the per-circuit percent failure rate is given per $10^3$ hours. The numbers noted on the curve refer to the number of circuits per chip. (b) Bipolar memory reliability, where the per-bit percent failure rate is given per $10^3$ hours and the numbers noted on the line refer to the number of bits per chip (K = 1024). (After Hsiao et al. 1981).

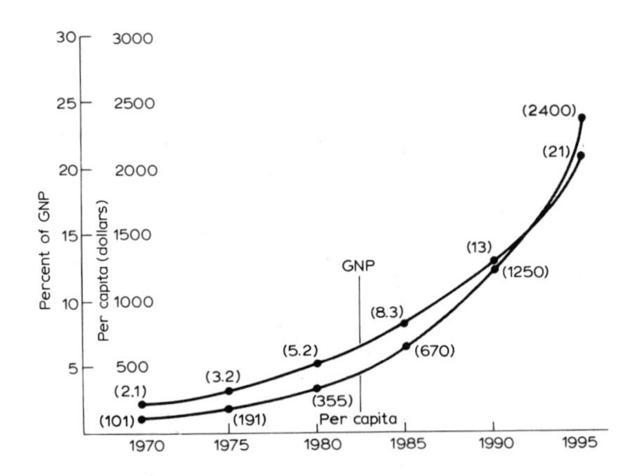


Fig. 4. Forecast of United States spending on data processing. Users of data processing equipment and services will spend an amount that will increase to 21% of the gross national product (GNP) by 1955. This amount will equal B2,400 per person. (After Gore and Stubbe 1979.)

Table 1: Some of the inventions which contributed to advances in office automation

| Date | Invention | Inventor | Nation |
|---|---|---|---|
| 1620 | Slide rule | Oughtred | English |
| 1642 | Adding machine | Pascal | French |
| 1809 | Paper machine | Dickinson | U.S. |
| 1823 | Calculating machine | Babbage | English |
| 1837 | Telegraph, magnetic | Morse | U.S. |
| 1868 | Typewriter | Soule, Glidden | U.S. |
| 1876 | Telephone | Bell | U.S.-Canada |
| 1879 | Cash register | Ritty | U.S. |
| 1884 | Punch card accounting | Hollerith | U.S. |
| 1884 | Pen, fountain | Waterman | U.S. |
| 1887 | Comptometer | Felt | U.S. |
| 1888 | Pen, ballpoint | Loud | U.S. |
| 1892 | Addressograph | Duncan | U.S. |
| 1899 | Tape recorder, magnetic | Poulsen | Danish |
| 1920 | Condenser microphone | Wente | U.S. |
| 1927 | Television, electronic | Farnsworth | U.S. |
| 1928 | Teletype | Morkrum, Kleischmidt | U.S. |
| 1938 | Xeroxography | Carlson | U.S. |
| 1939 | Computer, automatic sequence | Aiken, et al. | U.S. |
| 1947 | Transistor | Shockley, Brattain, Bardeen | U.S. |
| | Integrated circuits | | |
| | Word processors | | |
| | Advances in software and data-based management | | |

Bailey et al. (ref. 1) provided a critical review of the technology of office automation. In this regard a distinction is made between word processors and office information systems (OIS). Of OIS capabilities, Bailey et al. (ref. 1) indicated that:

> In particular, office automation aids office personnel in the preparation of documents, information management, and decision making. The types of documents to be prepared, processed, and managed by an OIS are memos, letters, reports, and typical business forms such as requisition and purchase order forms. The documents, still visually depicted in their regular format, are stored electronically in an OIS. The methods of capturing the information provided to the office workers produce documents with the standard headings and lettering appropriate to each particular form.
>
> The office worker interacts with the OIS through work stations. A work station is a programmable microcomputer equipped with office-oriented devices and software. The work station will enable the office workers to see and manipulate the electronic documents, communicate electronically with other work stations, access data base information, and stepwise direct the work station

to perform designated tasks on a one-time basis or automatically. Such a one-time task might be preparing a memo incorporating only sections of a previous memo and a previously prepared summary report.

An example of a task performed automatically by the work station might be searching an inventory data base to determine whether items listed on any purchase requisition are in stock or on order or whether they need to be special ordered. The work station operator could then fill out a blank purchase order, using the information found on the requisition form and in the inventory data base. Additional accesses to other system information could be made as the situation warranted. The completed purchase order form could then be transferred to another work station determined by the nature of further processing, such as managerial approval of the purchase.

In addition to electronic documents and the traditional data types, OISs will support video data such as pictures and graphs. Graphic packages are currently available and can be integrated into the OIS. Speech recognition would be a very desirable feature of an OIS, since some managers are opposed to keyboard interaction with the work station and prefer dictation and recordings in communicating with office personnel. Currently, speech recognition systems are in a very primitive stage of development. An experimental speech understanding system, Hearsay-II, recognizes utterances in a 1000-word vocabulary, with a correct interpretation rate of approximately 90%. Other alternatives to the keyboard are touch screens and light pens.

Communications between work stations can be instantaneously carried out by means of electronic mail or other electronic-type messages sent over telephone lines or by way of a satellite communication network. The electronic mail system will contain known addresses and other pertinent information such as telephone numbers. The sending and receiving of mail is handled automatically with little interaction by the work station operator. A "mailbox" is maintained for each work station, the contents of which may be surveyed upon request. Immediate notification of newly arrived mail is possible if desired, and an indicator light that shines when the mailbox is not empty is feasible.

The physical configuration of an OIS could be as simple as a single processing unit supporting several work stations or as complicated as a distributed network of large, interconnected computers, each of which supports a cluster of work stations. The degree to which each work station can operate independently of other work stations and network processing devices determines the degree to which the OIS is immune to hardware malfunctions and downtime. Each work station will have local data bases and data objects and indirect access to information local to another work station. The design of a particular OIS will determine which data bases are available to each work station at any given moment and how that information may be accessed and used.

Multiple tasks may be performed in parallel at individual work stations. This will occur most often when the office worker describes entire tasks or parts of tasks so that the procedures may be handled automatically by the work station. When the work station encounters an unanticipated discrepancy or situation, it will notify the office worker, describe the problem, and await further instructions.

In addition to the concurrent processing at individual work stations, parallel processing may occur when multiple work stations perform operations on a single transaction. For example, consider the purchase order that requires the independent approval of two managers. Once the purchase order is ready for the approvals, it would arrive simultaneously at the two managerial work stations. Each manager could then approve the purchase, put it into a "wait state" for future action, or block the purchase by denying approval. The main point is that the two approval processes do not necessarily have to be staged serially; instead they can be performed concurrently. If one manager should disapprove the purchase, the other could be notified, and the purchase order handled accordingly.

A prototype of office automation was provided by White (ref. 15). Twelve management work stations were installed at Citibank in 1976; and were intended to alleviate office workers' management and processing problems with the paperwork involved in customer transactions.

Even with the introduction of new technology to the offices, 1980 estimates indicated only about B2000 worth of equipment was available per U.S. office worker in comparison to B25,000 per worker in agriculture and B35,000 per worker in manufacturing.

4. HUMAN PERFORMANCE AND BEHAVIOR

Before one can fully appreciate how most effectively to interface office technology to humans, one must consider some basic human attributes that may affect this interaction:

a. Eleven percent of all office workers are left-handed; hence, the technology must be so designed that both left- and right-handed people can perform to their full potential.

b. One out of every 15 male office worker is color deficient; but only one out of every 10,000 female officer worker is color deficient. Red-green color deficiency is the most common. This has implications for the design and provision of color graphic displays to office workers.

c. In designing the job control of office work, consideration should be given to the fact that 10% of the labor force, in America, do not like work of any type; the remaining workers are split evenly between those who prefer to work at enriched jobs and are more satisfied and productive at it and those who prefer to work at simplified jobs at which they are more satisfied and productive (ref. 10). Hence when designing office jobs, workers should have a choice in performing either a simplified or an enriched job.

d. All furniture and equipment in the automated office should be adjustable so that people of all shapes and sizes can effectively and comfortably perform their tasks.

Two more factors to consider are the occupational stress and the social isolation of workers in automated offices.

a. Occupational Stress. The notion that immediate performance feedback accelerates the training process (ref. 4) and improves worker motivation (ref. 6) is well documented. Performance feedback has another useful function. Knight and Salvendy (ref. 7) demonstrated (Table 2) that as the performance feedback is made more exact, the worker's stress decreases. While performing exactly the same tasks, subject's stress levels were demonstrated to be nearly twice as high when no performance feedback was received as when exact performance feedback was given. Performance feedback can be utilized to manipulate worker's arousal and stress to maximize production output and/or physical and mental well-being (Figure 5).

Table 2. Effects of performance feedback on stress. (After Knight and Salvendy 1981.)

| Work condition | Stress index* |
|---|---|
| Self-paced | |
| No feedback | 100 |
| Cycle feedback | 90 |
| Time feedback | 86 |
| Combined time and cycle feedback | 57 |
| Machine-paced | 62 |

* A difference in the stress index of six or more units is statistically significant at 5% level.

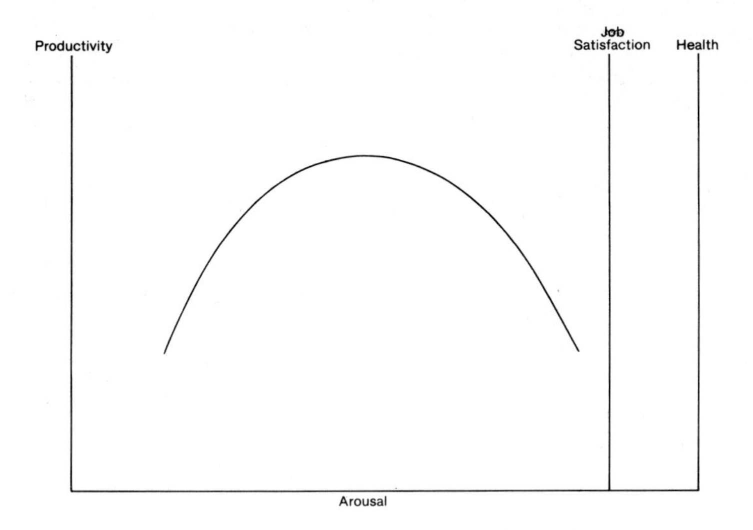


Fig. 5 Too much or too little arousal for task performance results in low productivity, job satisfaction and worker's health. There is an optimal region of arousal needed for maximizing productivity, job satisfaction and worker's health. This optimal region may vary depending on the criterion to be optimized.

However, effective stress management may also be influenced by the mental load involved in task performance. According to a study by Salvendy and Humphreys (ref. 11) in which either a computer or the subject controlled the work pace, it was revealed (Table 3) that for tasks requiring low mental load there were no differences in stress level between pacing modes; however, for the task requiring high mental load, the stress level was significantly higher for the self-paced than for the computer-paced work. These results are attributed to the notion that the computer keeps track of work output; whereas in self-paced tasks operators keep time which increases the mental load associated with task performance beyond that which is related to task content.

During a workshop following the International Conference on Machine-Pacing and Occupational Stress (ref. 12), it was suggested by the participating scientists that the most critical variable effecting stress on paced work is the length of the cycle time associated with a

Table 3. Impact of machine-paced and self-paced work on sinus arrhythmia scores. (After Salvendy and Humphreys 1979.)

| | Sinus arrhythmia | |
|---|---|---|
| | Mean | Standard deviation |
| Task with low perceptual level | | |
| Machine-paced | 0·66 | 0·15 |
| Self-paced | 0·69 | 0·13 |
| Task with high perceptual level | | |
| Machine-paced | 0·90 | 0·18 |
| Self-paced | 0·55 | 0·16 |

job. This notion was tested in a pilot experimentation (ref. 7) suggesting the possibility that for both machine-paced and self-paced tasks, stress decreases as cycle time increases. In this case, as in real-world industrial work situation, short and longer cycle times are confounded with simplified and enlarged jobs; hence, it is not feasible to assess whether it is the length of cycle or changes in the job content which cause this possible significant interaction.

Another variable, affecting occupational stress in human-computer interaction, is anticipation of systems delays. In this connection, one question is its impact on stress. Sharit and Salvendy (ref. 14) examined worker's stress in this connection. Table 4 indicates the results of the study and pinpoints that when the task performed prior to or after systems delay requires external attention (e.g., visual detection), the stress level during the anticipatory systems delay is significantly lower than for tasks requiring internal attention (e.g., arrhythmatic problems).

Together with the significant pacing effect, these results suggest that when a computer system is shared with more than one class of task activities, then from a human stress point of view, system delays should occur for those tasks that are self-paced (rather than computer-paced) and require external (rather than internal) attention.

Table 4. Acceleration and deceleration in heartbeats (per minute) as a function of job content, job design and the nature of financial incentive. Data after Sharit and Salvendy (1982).

| | No financial incentive | | Financial incentive | |
|---|---|---|---|---|
| | M/P | S/P | M/P | S/P |
| External attention | 66 | 63 | 68 | 62 |
| Internal attention | 71 | 69 | 75 | 72 |

Mean starting heartbeat 72 beats per minute.

b. Social Isolation. Perhaps the most significant impact of office automation is worker constraint and social isolation. In typical non-automated offices, workers move around the office to file and retrieve documents which are used at their office desk work. In contrast to this, workers in automated offices are more constrained since both the filing and retrieval of documents are stored in the computer memory which is accessible from the desk location of the worker.

In automated offices, workers interacting with computer systems are faced with systems delays which create uncertainty. This uncertainty causes stress and anxiety among these workers -- a situation which does not exist in conventional office work.

In many automated offices, the work rate of information processing and decision making is much higher than when the same tasks were done in a conventional office setting.

For these reasons, based on job evaluation principles, it has been argued in behalf of the Brotherhood of Utility Workers of New England that both Massachusetts Electric Company, and Naragansett Electric Company should pay the workers in automated offices 8- to 10-percent more than that received by workers doing the same tasks in the conventional office setting.

REFERENCES


1 A.D. Bailey, Jr., J. Gerlach, R.P. McAfee, and A.B. Whinston, Office Automation, Chapter 12.7 in Handbook of Industrial Engineering, G. Salvendy (Ed.), John Wiley & Sons, Inc., N.Y. (1982) pp. 12.71-12.7.20.
2 M.R. Gore, and J.W. Stubbe, Computers and Data Processing, New York: McGraw-Hill Book Company, 1979.
3 W.E. Harding, Semiconductor Manufacturing in IBM, 1957 to the Present: A Perspective, IBM Journal of Research and Development, 25 (1981) 647-658.
4 D.H. Holding, Human Skills, New York: John Wiley & Sons, 1981.
5 M.Y. Hsiao, Carter, W.C., Thomas, J.W., and W.R. Stringfellow, Reliability, Availability, and Serviceability of IBM Computer Systems: A Quarter Century of Progress, IBM Journal of Research and Development, 25 (1981) 453-465.
6 D.R. Ilgen, Fisher, C.D., and M.S. Taylor, Consequences of Individual Feedback on Behavior in Organization, Journal of Applied Psychology, 64 (1979) 349-371.
7 J.L. Knight, and G. Salvendy, Effects of Task Feedback and Stringency of External Pacing on Mental Load and Work Performance, Ergonomics, 24 (1981) 757-764.
8 P.F. Olsen, and R.J. Orrange, Real-Time Systems for Federal Applications: A Review of Significant Technological Developments, IBM Journal of Research and Development, 25 (1981) 405-416.
9 J.R. Opel, Message from the President and Chief Executive Officer, IBM Journal of Research and Development (25th Anniversary Issue), 25 (1981) front page.
10 G. Salvendy, An Industrial Engineering Dilemma: Simplified versus Enlarged Jobs, In Production and Industrial Systems: Future Development and the Role of Industrial and Production Engineering, in R. Muramatsu and N.A. Dudley (Eds.), London: Taylor & Francis, Ltd., 1978.
11 G. Salvendy, and A.P. Humphreys, Effects of Personality, Perceptual Difficulty and Pacing of a Task on Productivity Job Satisfaction and Psychological Stress, Perceptual and Motor Skills, 49 (1979) 219-222.
12 G. Salvendy, and M.J. Smith, (Eds.), Machine Pacing and Occupational Stress, London: Taylor & Francis, Ltd., 1981.
13 G. Salvendy, Human-Computer Communications with Special Reference to Technological Developments, Occupational Stress and Educational Needs, Ergonomics, 25 (1982) 435-447.
14 J. Sharit, and G. Salvendy, Occupational Stress: Review and Reappraisal, Human Factors, 24 (1982) 129-162.
15 R.B. White, A Prototype for the Automated Office, Datamation, (1977) 83-90.

# WORK ENVIRONMENT ISSUES OF SWEDISH OFFICE WORKERS: A UNION PERSPECTIVE

OLOV OSTBERG
Central Organization of Salaried Employees
in Sweden (TCO)
Box 5252
S-10245 Stockholm, Sweden

## IDENTIFYING TROUBLESOME FACTORS

In the process of creating a Work Environment Laboratory at the Stockholm University of Technology, Magnus (ref. #1) helped to set the priorities of the laboratory by studying troublesome jobs. A representative sample of Swedish worksites were visited; local safety and health, management and union representatives were instrumental in locating such troublesome jobs, that is, jobs that raised concern owing to safety and health problems, high turnover rates, staffing difficulties, complaints from the workers, uneven production output, and so forth. The next step was to identify the factors most likely to cause/contribute to the troublesome nature of these jobs. In order of importance, the contributing job factors are shown in Table 1.

TABLE 1
Job factors perceived as contributing to the troublesome nature of jobs. The factors are listed in order of importance (ref. #1).

1. Taxing static muscular workload
2. High sensory demands
3. Machine-paced time patterns
4. Taxing peak muscular workload
5. Noise
6. Gas, dust, fumes
7. Taxing dynamic muscular workload
8. Adverse temperature conditions
9. Vibration
10. Unsatisfactory lighting

It may come as a surprise that traditional hard work environment factors such as heavy physical work and chemical hazards did not rank higher among the factors of Table 1. Apparently many sedentary jobs could well qualify as having the ingredients of troublesome jobs. These findings were reported in 1970, and it would be interesting to have the study repeated. The Swedish Confederation of Trade Unions (LO, a blue-collar organization), has provided information to this effect, but only based on workers' own perceptions of their work environment. In two large questionnaire investigations, in the years 1969 (ref. #2) and 1980 (ref. #3), identical questions were asked about the presence of adverse work environment conditions. The most definite finding was that the stress and related "soft" ergonomic factors had worsened. In their ensuing action program on psychological and social health risks (ref. #4), LO urged white-collar workers to take heed of these experiences of blue-collar workers in order to prevent the worsening trend from spreading to the office world. Unfortunately, the official Swedish statistics indicate that not only are factory and office workers alike in terms of their perceptions of the new health risks, but a breakdown of the statistics from 1980 (ref. #5) actually shows ergonomic factors to be the predominant causation of occupational diseases (see Table 2).

TABLE 2
Breakdown of the 1980 official Swedish statistics on occupational diseases (ref. #5).

| Factors of Occupational Diseases | Incidence (%) |
|---|---|
| Ergonomic factors[a] | 52.9% |
| Chemical factors | 22.1% |
| Noise | 12.1% |
| Biological hazards | 3.2% |
| Other causes | 9.5% |

[a]This causation category contains occupational diseases of a musculo-skeletal nature, caused by unsuitable work postures, work movements, and workloads.

To a high degree, the occupational disease category ergonomic factors (Table 2), combines and overlaps with the top three troublesome job factors (Table 1): static muscular workload (for example, bad work posture), high sensory demands (for example, difficulties seeing), and machine-paced time patterns (for example, delayed machine response times). Furthermore, those are also the factors that best describe many of today's office jobs with video display terminals (VDTs). This situation has resulted in the following U.S. recommendation (ref. #6):

> Based on our concern about potential chronic effects on the visual system and musculature and prolonged psychological distress, we recommend the following work-rest breaks for VDT operators:
>
> 1. A 15-minute work-rest break should be taken after two hours of continuous VDT work for operators under moderate visual demands and/or moderate work load.

2. A 15-minute work-rest break should be taken after one hour of continuous VDT work for operators under high visual demands, high work load and/or those engaged in repetitive work tasks.

It is interesting to compare these recommendations with the 1964 Japanese national safety and health ordinance on key-punch operations. In 1960, it was observed that key-punch operators in Japanese banking and stock market businesses displayed impairments in the hands/fingers, and shoulders/arms after the introduction of computer systems into office work. The problems have been substantially reduced since it was decided that:

- Key-punch work should be limited to 300 minutes per working day.
- One continuous working period for key-punch work should not exceed 60 minutes.
- Keystrokes should not exceed 40,000 strokes per day.

## RESPONSES TO NEGATIVE FACTORS IN THE OFFICE ENVIRONMENT

### By Workers and Unions

Office workers had, of course, felt the presence of paced and regimented production systems, boring and routinized jobs, unsuitable work postures, and strenuous display viewing conditions, long before the research community became interested in these topics. TCO, the Central Organization of Salaried Employees in Sweden (a white-collar organization), has for many years actively attempted to reduce the time between workers' awareness of work environment problems and scientists' acknowledgment of the very same problems. By and large the scientific community is sensitive to well founded early warning signals, but now and then this backfires in that scientists set out to prove that the signals are not trustworthy (ref. #7).

It is the view of TCO that research and development (R&D) is a function on which trade unions have the right to exercise responsible influence. Getting involved in R&D is thought to be a worthwhile investment of sparse union resources. Society does not bestow this authority on unions. It is rather the result of the high degree of unionization and the concommitant power position. In the Swedish office domain, approximately 85% of the employees belong to a union--even professionals and management personnel. TCO's engagement in R&D for a better office work environment has been described elsewhere (ref. #8,9,10), and can be seen as a natural development in the struggle for a voice in the design, implementation and operation of office automation systems.

The changing office world in general, and office automation in particular, has led to the notion of <u>office factories</u>. In response to this altered perception, office workers make reference to their working conditions in terms previously heard only from factory workers. The job factors of troublesome jobs and the statistics of occupational diseases now seem to be equally applicable to office and factory work. Similar signals came from nurses, journalists, teachers, police officers, and other categories of salaried employees. To explore these emerging problems, TCO eventually sent out a questionnaire to 12,000 of its approximately one million members. Eighty-five percent of the questionnaires were satisfactorily completed and returned. TCO used the results of the questionnaire (ref. #11) in the legislative process that led to the adoption of a new Swedish Work Environment Act, 1977. The following are the new key points laid out in the Act:

- The working environment shall be satisfactory in relation to the nature of the work concerned and to social and technical progress in society at large.

- Working conditions shall be adapted to human physical and mental aptitudes.

- Efforts shall be made to arrange work in such a way that the employee can influence his own working situation.

A new Work Environment Act will not automatically result in better working environments. With regard to the reference to "social and technical progress," it should be noted that the Act was adopted in a period of relative affluence, when social and technical progress were going hand in hand. This is no longer an unchallenged fact, and accordingly the employers are now anxiously pointing to forecasts of economic recession trends and demanding concessions from the employees. Unions, therefore, find it increasingly important to keep the work environment discussion open to avoid being squeezed by the rapid technological development, economic recessions, and a worldwide political movement advocating deregulation of safety and health.

By Researchers

One way that the TCO unions have kept a continued discussion open, and underlined demands, has been to make follow-up investigations based on the TCO work environment questionnaire. Such research reports have now been published for the unions of flight attendants, police officers, teachers, journalists, and municipal employees. These investigations have been carried out by independent researchers from the Swedish scientific community. TCO has also benefited from a thorough re-analysis of the TCO work environment questionnaire carried out by an international research team.

The TCO questionnaire data were analyzed with regard to possible connections between job characteristics and ill-health, as reported by the questionnaire respondents. The theoretical framework was designed by Robert A. Karasek, who, with his colleagues, reported the results (ref. #12,13). A full account of the model, and a discussion of its generality among blue- and white-collar workers both in Sweden and the U.S., has been published in the NIOSH-sponsored report (ref. #14) from which Figure 1 is adapted.

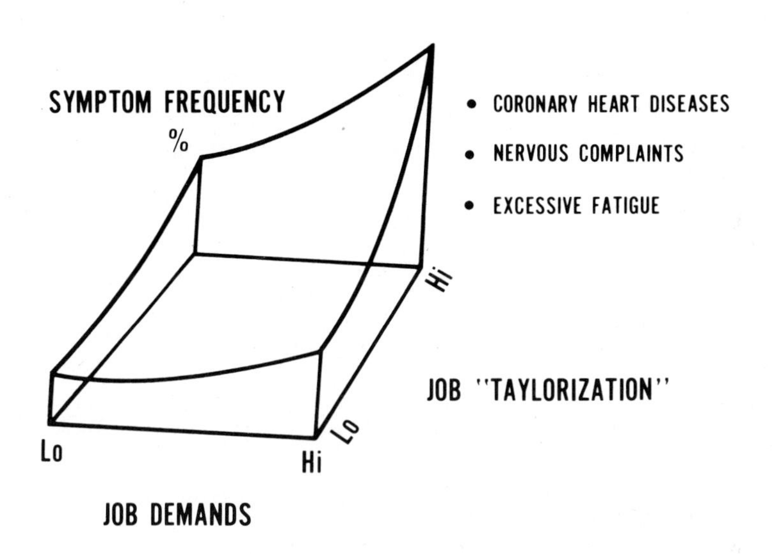


Figure 1. Summarized relationship between job characteristics and ill-health symptoms. Job demands denotes mental and physical work loads. Job Taylorization denotes regimentation, pacing, routinization, lack of decision latitude, etc. (From Karasek and others, ref. #14)

Karasek and his associates have developed a model which, applied to TCO data, supports and gives scientific credibility to the long-held TCO claims that there is a relationship between job characteristics and ill-health among salaried employees. With reference to Figure 1, job demands is a collective term for mental and physical workloads of various sorts, and job Taylorization denotes such negative job aspects as regimentation, pacing, lack of skill discretion, and lack of decision latitude. If this model of analysis is applied to the TCO data, the results clearly show that high job demands result in increased frequency of ill-health symptoms, and, more important, that this relationship is amplified many times for salaried employee groups whose jobs score high on the Taylorization scale.

WHAT IS TAYLORIZATION?

The term Taylorization may need to be explained. In Sweden this term has a negative connotation. Several Swedish unions have formulated new technology action programs that specifically mention that a Tayloristic job organization will not be accepted (ref. #8). Even the Swedish employers have declared that Tayloristic production systems should preferably not be employed in tomorrow's factories. This declaration is to be found in a policy book on production management published by the Swedish Employers' Confederation (SAF), and from which the following description of the Taylorization of Sweden is borrowed (ref. #15):

> "The man in the planning room, whose speciality under scientific management is planning ahead, invariably finds that the work can be done better and more economically by subdivision of the labor; each act of each mechanic, for example, should be preceded by various preparatory acts done by other men."
>
> The above is from the book The Principles of Scientific Management which was first published in the U.S. in 1911 and in Sweden in 1913. The author was Frederick W. Taylor, who by this time had become famous throughout the world for his new ideas on the organization and management of work in American companies, summarized in the expression he invented--"Scientific Management." Taylor's ideas spread quickly in Swedish industry, as they did in many other industrialized countries. Few persons have had such a great impact on the practical application of organizational theory as Taylor. The division of tasks, specialization, systematic analysis of the work cycle, pre-decided time limitations, careful follow-up--all this became the basis for a dramatic injection of efficiency in American industry, and in Europe we also were quick to accept the new principles.

Now the Swedish Employers' Confederation is in the process of abandoning the school of Taylor, which is easier said than done. To be sure, since Taylor's days, many new theories about how to organize, manage, and control work in companies have emerged, but most of them have been constructed on essentially the same foundation. In this connection it is worth noting that the U.S. management guru and Wall Street Journal columnist Peter F. Drucker considers automation to be a logical extension of Taylor's Scientific Management (ref. #16). It is actually a widespread Swedish apprehension that computer products and systems imported from the U.S. will somehow be accompanied by influx of Tayloristic management principles. Humanized, user-friendly, and ergonomically designed automation--it all sounds good, but the unions have learned by experience to be suspicious. Humanization is sometimes synonymous with the removal of workers as a result of automation.

How can a new system be user-friendly when the users (the workers) were denied any voice in the design, implementation, and operation of the system? Lenin introduced Scientific Management into the U.S.S.R. under the name of ergonomics, and both in Europe (ergonomics) and the U.S. (human factors) this link is still visible in that the discipline promotes the comfort and efficiency of workers.

## WORKING TOWARD THE FUTURE

The word management in the concept Scientific Management came into English from the Italian word "maneggiare" which means "to handle and train horses." One would hope that today's office management has not inherited the view that scientific tools and computer mediated strategies are needed to prevent the office workers from horsing around. TCO urges that the factory of the past is not used as a model for the office of the future.

The word scientific in the concept Scientific Management has a positive connotation, as has R&D in general. TCO acknowledges the importance of scientifically founded societal decisions and can capitalize on and draw strength from research results such as those shown in Figure 1. The current TCO Programme of Action (ref. #17) demands that the Swedish national R&D budget shall be increased from 2% to 3% of the gross national product, and also lists specific R&D areas that are particularly relevant to the office work environment. However, this does not mean that actions cannot be initiated in the absence of a scientific platform. In a report entitled The Elusive Working Environment (ref. #18), TCO actually takes the position that actions to improve the psychological and social working environment have been suppressed by too much patience and humility with respect to SCIENCE. We must not wait for the researchers to cast our problems in their scientific molds, says the report; instead, we must address the problems when they become recognized at the local level. For example, stress is usually the result of management style combined with existing overt/covert wage policy, staffing policy, employment policy, resource allocation policy, decision latitude policy, etc. Stress problems will be addressed accordingly, and low priority will be given to superficially and/or individually oriented programs on stress awareness and relaxation, screening for optimal person-environment fit, drugs and alcohol programs, physical fitness programs, and social support promotion.

### Some practical concerns

Recent examples of superficially and individually oriented programs are found in a booklet on Work Positions and Work Techniques published by the medical department of IBM of Sweden (ref. #19). The booklet's advice and instructions are sound in a narrow perspective, but perhaps oversimplified.

For example, performing eye exercises now and then cannot be the sole remedy for eyestrain in VDT work. Totally unacceptable, not to say comic, advice is given in the IBM-financed scientific paper by Campbell and Durden (ref. #20). They advocate that a modern and correctly used VDT should not cause any eyestrain. They suggest that VDT operators' complaints of eyestrain can, to a large extent, be ascribed to boredom; they recommend that these VDT operators should work faster so that they have no time to be bored or feel any eyestrain. As pointed out by Stark (ref. #21), even though eyestrain is not a well-defined psychological or clinical entity, this cannot negate the fact that many highly motivated VDT users suffer from ocular discomfort and visual fatigue beyond that appropriate to a normal workplace. An example of a nonsuperficial and nonindividual approach is the new Swedish health and safety ordinance (ref. #22), in which is established that as of January 1, 1984, the viewing conditions (in VDT operations, for example) shall be such that taxing work postures will not be called for, and that the workers as a second step shall be provided with visual aids, tested for the individual worker/task.

Work environment issues and the formulation of ordinances and standards are political in the sense that they are linked to labor relations issues. A look at the current goals of the U.S. National Institute for Occupational Safety and Health (NIOSH) may clarify this point. These are the NIOSH goals:

- To prevent premature mortality of workers,
- To reduce unnecessary morbidity of workers,
- To improve the quality of life of workers.

These are goals that can also be agreed upon by employees and employers, but the further down the list we go, the more the views diverge in a practical situation. For example, employees' complaints about visual and musculoskeletal strain connected with VDT work are often met with the employer's argument, that there is really no ground for complaints because VDT work is not much different from typewriting, and typewriters have been in the office forever. Researchers sometimes take over this argument and try to prove that when conditions are otherwise equivalent, an electroluminiscent display (VDT) can be read with no more eyestrain than when reading a hardcopy (paper). To this can be said that it is futile in a laboratory set-up to make conditions "otherwise equal," because these two types of technology are fundamentally different. Furthermore, it is invalid to state that in terms of eye effects there are no differences between proofreading with hardcopy vs. with a VDT from IBM (ref. #23), when (1) the hardcopy reading was with fewer errors and 30% faster, and (2) the subjects in both cases became temporarily nearsighted as a result of the strenuous visual task.

Figure 2. Telephone operators in Malmo, Sweden, 1896. (Photo: Televerket)

Figure 3. Telephone operators in Stockholm, Sweden, 1982. (Photo: Televerket)

Traditional office work and computerized office work are very different, yet similar. The photographs in Figures 2 and 3 show Swedish telephone operators from the years 1896 and 1982 respectively; verbal descriptions of the 1982 work environment are available from management (ref. #24), union (ref. #25), and research (ref. #26,27) perspectives. The modern telephone operator's office looks nicely decorated, ergonomically well designed, adjustable, relaxed, and comfortable. But note that a supervisor is still overseeing the production. Not seen in this photograph is that the agreement between management and unions prohibits the supervisor from obtaining performance data from individual operators--only for the production group--and that the operators switch between VDT and non-VDT work every 1 1/2 hours. Also not visible in photographs of automated offices is the new and impersonal bureaucracy brought about by the computer system's need for strict information standards, sequences, and hierarchies, which means that the computer-mediated work (ref. #28) is abstract and formal and requires conceptual skills rather than social skills and direct experience.

So far the computerized office has not become the paperless office. The paper volume actually tends to increase in proportion to the volume of office automation equipment. Of particular concern is the amount of chemicals introduced into offices. Xerographic copying/printing machines from IBM have been shown to emit a mutagenic chemical named trintrofluorenon (TNF), which has led IBM to issue a firm recommendation on the use gloves for service personnel (for a discussion, see ref. #29). Copying machines from Xerox have also been shown to emit mutagenic chemicals. Scientists from Xerox (ref. #30), ascribed this to impurities in certain brands of carbon black and toners, which were then withdrawn from the market. The products that remained on the store shelves in Sweden after the withdrawal were reduced in price; one governmental agency took advantage of this sell-out and hoarded several years' supply of mutagenic carbon black. The Swedish National Board of Occupational Safety and Health refused to say which agency did this. The Board also protected industry during TCO's intense drive to bring about safety and health measures affecting carbonless copying paper, which has caused skin, eye, and mucous membrane problems for many office workers (ref. #31). The Board testified that some types of carbonless copying paper ("high problem paper") cause more problems than other types ("low problem paper"), but refused to inform TCO which paper is which (ref. #32) and instead appealed to the paper producers to adopt self-censoring. TCO has not despaired; its pamphlet on carbonless copying paper (ref. #33) is used all over Sweden and has been instrumental in reducing or eliminating the use of such paper at many workplaces.

The symptoms associated with handling large amounts of some brands of carbonless copying paper resemble the so-called "VDT dermatitis" (ref. #34,35). Perhaps there are even synergistic effects where this paper constitutes the input documents in VDT data entry operations. One common denominator could be the low humidity in Scandinavian offices in wintertime, allowing static electric fields to build up and increasing the capacity of the air to hold microscopic dust particles while at the same time making the operators' mucouse membranes more susceptible to irritants. In Denmark, Dreyer and others (ref. #36) have found dry air to be a common denominator in situations where VDT and non-VDT operators complain about strained and irritated eyes. The mechanism is believed to be that the tear film of the eye easily breaks up in dry air.

Dry air is also one of the components in the sick building syndrome (ref. #37). In Sweden, these problems were first referred to as the "kindergarten disease" because they were first reported among children and attendants of such institutions. The problems were severe enough for the Factory Inspectorate to close down several kindergartens. It took some time, though, before the official safety and health institutions addressed the problems, which were first looked upon as artifacts related to "problem persons" and to "neurotic women in their transition age." Such an attitude is naturally met with dismay from employees, and although TCO does not contest the existence of psychological effects of indoor air pollution (ref. #38) the organization strongly objects to using the concept of "mass psychogenic illness" (ref. #39) as a null hypothesis whenever unions are raising questions about safety and health in the office.

Employees' worries about unknown work environment risks should be taken seriously and, as an example, the TCO policy in the booklet VDT Work (ref. #40) is that pregnant VDT operators should have the right to temporarily switch to non-VDT work if the miscarriage debate (ref. #41) has worried them. Such worry is justified considering that VDTs are operating on high voltage and that medical journals are reporting connections between high voltage and childhood cancer (ref. #42) and leukemia (ref. #43), and warns about harmful radiation from older TV sets (ref. #44). It is also known that the static electric field around a VDT can provoke facial rash in sensitive operators, under certain conditions (ref. #45), so there are indeed good ground for worry among VDT operators. The results from a study of pregnancy problems among hospital employees are also relevant (ref. #46). The study was initiated as a result of worries about a possible connection between miscarriages, and other pregnancy problems, and the handling of chemicals in the University of Gothenburg hospital laboratories. No such relationship was found, nor could the cigarette smoking pattern explain the pregnancy problems, although birth weights were lower when the mother smoked during preganancy. Shift work during pregnancy was the one and only explanatory factor. The relative risk for pregnancy problems was signficantly higher in those having night or shift work (RR 3.2, 95% confidence 1.4-7.5), a finding that suggests the importance of looking into diurnal rhythm, stress, and other psychological factors in the further research on miscarriages among VDT operators.

Larger Issues

In concluding the present paper on work environment issues as seen by Swedish office workers, it is TCO's experience that office automation does not automatically result in improved working conditions. The outlook for the future is a bit pessimistic as evidence emerges that, according to Levin and Rumberger (ref. #47): High technology and automation will probably (1) lower job skills, (2) eliminate more jobs than they create, (3) simplify and routinize work tasks, (4) reduce worker individuality, (5) reduce the areas of worker judgment, and (6) reduce the possibility of job-related social contacts.

To turn the tide, all parties involved in office automation projects ought to sign and live up to the declaration that: Employees have a legitimate right to ask for protection of their (1) employment, (2) safety and health, (3) income, (4) craft-skill, (5) growth and promotion potential, (6) decision latitude, and (7) human dignity.

Each nation, of course, should address its problems in its own idiosyncratic way, and this paper is not suggesting that the solutions favored by unionized Swedish office workers will be effective elsewhere. According to an authoritative U.S. spokesperson (ref. #48), the Scandinavian countries rank highest among the industrialized countries when it comes to end-user (worker) participation in systems design. At the other extreme of the continuum is the U.S., which probably has the lowest level of participative management and participative information systems design, and is characterized by traditional adversary relations between management and labor. This is discouraging, especially as the U.S. is in the lead when it comes to producing hardware and software for the office automation, but according to the same spokesperson:

> In the U.S., it is my prediction that the participative approach will eventually displace traditional system design by the end of this century, but with features unique to American culture and the American organizational setting.

It is the experience in the Scandinavian countries that such worker participation will result in better working conditions as well as a more effective use of productive resources in the company (ref. #49).

ACKNOWLEDGMENT

The preparation of this paper took place while the author was on leave from TCO, Stockholm, Sweden, to be a visiting research scientist with NIOSH, Cincinnati, OH, U.S.A. Stimulation and encouragement is acknowledged from P.E. Boivie, M. Breidensjo and I. Wahlund, former TCO colleagues, and B. Amick and M. Smith, present NIOSH colleagues.

REFERENCES

1 P. Magnus, Investigation of Troublesome Jobs, The Swedish Board for Technical Development (STU), Stockholm, 1970. (In Swedish)
2 Risks in the Job: The LO Questionnaire Investigation, Swedish Confederation of Trade Unions (LO), Stockholm, 1970. (In Swedish)
3 What is Happening With the Working Environment?, Swedish Confederation of Trade Unions (LO), Stockholm, 1981. (In Swedish)
4 The LO Action Programme on Psychological and Social Health Risks, Swedish Confederation of Trade Unions (LO), Stockholm, 1979. (In Swedish)
5 A. Kilbom, Occupational Disorders of the Musculoskeletal System. Newsletter of the Swedish National Board of Occupational Safety and Health, 1 (1983) 6-7.
6 Potential Health Hazards of Video Display Terminals, U.S. Department of Health and Human Services, Washington, D.C., 1981. (DHHS/NIOSH 81-129)
7 Video Displays, Work, and Vision, National Academy Press, Washington, D.C., 1983.
8 O. Ostberg, The Empirics of Specialization and Division of Labor Among Swedish Salaried Employees: A Trade Union View on the Technical Development. Proceedings of the International Symposium on Division of Labor, Specialization and Technical Development, Linkoping, Sweden, 1982.
9 O. Ostberg, Office computerization: research then and now, in D. Marschall and J. Gregory (Eds.), Office Automation: Jekyll or Hyde?, Working Women Education Fund, Cleveland, 1983, 127-142.
10 P. E. Boivie and O. Ostberg, Programme of Data Policy for the Central Organization of Salaried Employees in Sweden (TCO), in U. Briefs, C. Ciborra and L. Schneider (Eds.), System Design for, with and by the Users, North Holland, Amsterdam, 1983, 289-292.
11 I. Wahlund and G. Nerell, Work Environment of White Collar Workers: Work, Health and Well Being, Central Organization of Salaried Employees in Sweden (TCO), Stockholm, 1976.
12 R.A. Karasek, J. Lindell and B. Gardell, Patterns of Health Association with Job and Non-job Stressors for Swedish White Collar Workers, Department of Psychology, University of Stockholm, 1981. (Unpublished report)
13 B. Gardell, The Salaried Employees' Work Environments, Stockholm University Department of Psychology, Stockholm, 1979. (In Swedish, preliminary report)
14 R.A. Karasek, J. Schwartz and T. Theorell, Job Characteristics, Occupation, and Coronary Heart Disease, U.S. Department of Health and Human Services, Washington, D.C. 1982. (NIOSH grant report)
15 S. Aguren and J. Edgren, New Factories: Job Design through Factory Planning in Sweden, Swedish Employers' Confederation (SAF), Stockholm, 1980.
16 P.F. Drucker, Technology and society in the twentieth centruy, in M. Kranzenberg and C.W. Pursell Jr. (Eds.), Technology in Western Civilization, Oxford University Press, New York, 1967.

17 The TCO Programme of Action 1983-1985, Central Organization of Salaried Employees in Sweden (TCO), Stockholm, 1983.
18 The Elusive Working Environment: On the Psychological and Social Consequences of the Working Environment, Central Organization of Salaried Employees in Sweden (TCO), Stockholm, 1982. (In Swedish)
19 Work Positions and Work Techniques. Advice and Instruction, IBM Svenska AB, Stockholm, 1982.
20 F.W. Campbell and K. Durden, The visual display terminal issue: consideration of its physiological, psychological and clinical background. Ophthalmic and Physiological Optics, 3 (1983) 175-192.
21 L.W. Stark, Dissent. In Video Displays, Work, and Vision, National Academy Press, Washington, D.C. 1983, 235-236. (Dissent from panel report)
22 Work Postures and Work Movements. Ordinance of the Swedish National Board of Occupational Safety and Health, AFS 1983:6. (In Swedish)
23. J.D. Gould and N. Grischkowsky, Doing the Same Work With Hardcopy and With Cathode Ray Tube (CRT) Computer Terminals. IBM Research Report, RC 9849, 1983.
24. B. Tornqvist and S. Hermansson, Automation and the quality of work life at the Swedish Telephone Company: a management view, in D. Marschall and J. Gregory (Eds.), Office Automation: Jekyll or Hyde?, Working Women Education Fund, Cleveland, 1983, 79-83.
25. B. Westman, Automation and the quality of work life at the Swedish Telephone Company: a union view, in D. Marschall and J. Gregory (Eds.), Office Automation: Jekyll or Hyde?, Working Women Education Fund, Cleveland, 1983, 84-88.
26. A-Ch. Blomkvist, G. Lindgren and O. Ostberg, Accommodation Measurements With a Laser Optometer: Part of an Experiment on Visual Fatigue at the Swedish Telecommunications Services, University of Lulea Technical Reports, No. 78T, 1980. (In Swedish)
27. E. Gunnarsson and I. Soderberg, Eye strain resulting from VDT work at the Swedish Telecommunications Administration. Applied Ergonomics, 14 (1983) 61-69.
28 S. Zuboff, New worlds of computer-mediated work. Harvard Business Review, 60:5 (1982) 142-152.
29 J Makower, Office Hazards: How Your Job Can Make You Sick, Tilden Press, Washington, D.C., 1981.
30 H.S. Rosenkranz, E.C. McCoy, D.R. Sanders, M. Butler, D.K. Kiriazides and R. Mermelstein, Nitropyrenes: isolation, identification, and reduction of mutagenic impurities in carbon black and toners. Science, 209 (1980) 1039-1043.
31 T Menne, G. Asnaes and N. Hjorth, Skin and mucous membrane problems from "No Carbon Required" paper. Contact Dermatitis, 7 (1981) 72-76.
32 C.J. Gothe, I. Jeansson, A. Lindblom and D. Norback, Carbonless Copy Papers and Health Effects. Opuscula Medica (Stockholm), Suppl. LVI, 1981.
33 Carbonless Copying Paper: Facts and Problems, Central Organization of Salaried Employees in Sweden (TCO), Stockholm, 1982. (In Swedish)
34 V. Linden and S. Rolfsen, Video computer terminals and occupational dermatitis. Scandinavian Journal of Work, Environment, and Health, 8 (1982) 62-64.
35 A Nielsen, Facial rash in visual display operators. Contact Dermatitis, 8 (1982) 25-28.
36 V. Dreyer, S. Jensen, V. Pedersen, E. Petersen and A. Richter. The Working Environment at Visual Display Units: A Field Study, European Foundation for the Improvement of Living and Working Conditions, Dublin, 1983.
37 J. Messite and D.B. Baker, Occupational health problems--a mixed bag, in B.G.F. Cohen (Ed.), Human Aspects in Office Automation, Elsevier, Amsterdam, 1984.
38 M.J. Colligan, The psychological effects of indoor air pollution. Bulletin of the New York Academy of Medicine, Second Series, 57 (1981) 1014-1026.

39 M.J. Colligan and L.R. Murphy, Mass Psychogenic Illness in Organizations: an overview. Journal of Occupational Psychology, 52 (1979) 77-90.

40 VDT Work--The Correct Way, Central Organization of Salaried Employees in Sweden (TCO), Stockholm, 1983. (In Swedish)

41 L. Slesin and M. Zybko, Video Display Terminals: Health and Safety, Microwave News, New York, 1983.

42 N. Wertheimer and E. Leeper, Electric wiring configurations and childhood cancer. American Journal of Epidemiology, 109 (1979) 273-284.

43 S. Milham Jr., Mortality from leukemia in workers exposed to electrical and magnetic fields. New England Journal of Medicine, 307 (1982) 249.

44 S.D. Savic, More on radiation hazard of video screens. New England Journal of Medicine, 308 (1983) 458.

45 W.C. Olsen, Electric Field Enhanced Aerosol Exposure in Visual Display Unit Environments. The Chr. Michelsen Institute, Bergen, Norway, 1981.

46 G. Axelsson, C. Lutz and R. Rylander, Exposure to Solvents and Pregnancy Outcome Among University Laboratory Employees, University of Gothenburg Department of Environmental Hygiene, Gothenburg, 1982 (Report unpublished).

47 H.M. Levin and R.W. Rumberger, The Educational Implications of High Technology, Stanford University Institute for Research on Educational Finance and Governance, Stanford, 1983.

48 H. Sackman, Problems and promise of participative information system design, in U. Briefs, C. Ciborra and L. Schneider (Eds.), System Design for, with and by the User, North Holland, Amsterdam, 1983.

49 B. Gardell, Worker Participation and Autonomy: A Multi-Level Approach to Democracy at the Workplace, University of Stockholm Department of Psychology, Stockholm, 1982.

# SOME ISSUES SURROUNDING THE DESIGN OF ERGONOMIC OFFICE CHAIRS

MARVIN J. DAINOFF
Psychology Department
104 Benton
Oxford, Ohio

## INTRODUCTION

> ......let it not be assumed that bad postures are experienced only by machine operators. The requirement to stand all day set by many jobs is unequivocally a bad posture. So, too, is the requirement to sit all day. Indeed, it is arguable that the latter is worse than the former. The problems of motor vehicle drivers are classic in terms of restricted posture and the problems of office workers are likely to follow them into the literature if some of the office automation on the horizon becomes widespread.
>
> E. N. Corlett (ref. 1)

A converging line of evidence indeed suggests that postural difficulties associated with work at video display terminals have the potential for becoming an area of major concern. Dainoff's (ref. 2) review of the pre-1980 literature documents several indications of musculoskeletal problems associated with VDT working conditions. Sauter et al. (ref. 3) found VDT workers to have higher levels of reported musculoskeletal disturbances than did comparable non-VDT workers; correlational analyses showed these to be related to chair and workstation configuration. Recent field work by Grandjean and his colleagues (ref. 4) demonstrates a relationship between preferred dimensions of VDT workstation configuration and levels of musculoskeletal complaints.

In a controlled laboratory investigation conducted by the present author and his colleagues (ref. 5), a global comparison between two ergonomic extremes of VDT workplace design yielded fewer musculoskeletal complaints when participants worked a three-hour data-entry shift in the "good" workstation, as compared with a comparable work period in the "poor" workstation. In addition, participants yielded higher levels of work performance (increased keystroke rates) when in the "good" workstation. Given these findings, it is of interest to examine some salient ergonomic attributes which differentiated the "good" from the "poor" workstations. (For purposes of this paper only postural issues will be considered. Visual factors (lighting and glare) were manipulated in the experiment but did not appear to have any impact on the dependent variables.) In the good workstation, ergonomic recommendations from a highly quoted source (ref. 6) were followed in designing the environment. Specifically, participants were seated at a bi-level VDT workstand which was adjusted for the participants such that elbows were located approximately at the level of the home row of the keyboard, and viewing angle of VDT screen center was approximately 20 degrees. A copy holder was located between keyboard and screen. Participants were seated in a five-star based gas-cylinder height-adjustable chair, with independently adjustable backrest. The lower portion of the backrest contained a lumbar support pad.

In the poor workstation, an adjustable terminal base was also utilized, but this equipment was deliberately maladjusted in order to simulate poor working conditions. The keyboard support was moved upward so that forearm angle approximated 45 degrees, and screen support was moved downwards so that viewing angle was greater than 30 degrees. No copy holder was available; copy rested on the desk next to the keyboard. A standard office chair (nonadjustable without tools) with no lumbar support pad was employed. Chair height was set to a standard 47 cm.

Given the above results, along with one other published confirmation by Springer (ref. 7), and a number of personal communications of similar unpublished findings, it would seem that one could feel confident in recommending adjustable ergonomic terminal stands and chairs as effective additions to the electronic office workplace in terms of reduction of health complaints as well as increasing productivity. However, while experts in the area agree on the general concept of adjustability/flexibility, there appears to be considerable disagreement on specific details of implementation of this concept. This disagreement is seen particularly with regard to chair configuration. Accordingly, the purpose of the following discussion is to examine, from the perspective of the nonspecialist user of ergonomic recommendations, some issues regarding office seating.

## STANDARD RECOMMENDATIONS

E. Kroemer is seen as one of the most influential authorities in the area of ergonomic seating, and it will be useful to rely on his writings to assess the current state-of-the-art in this area. Kroemer and Robinette (ref. 8), in a review of the European literature, laid out some of the fundamental issues in seating design. They discuss the tendency of the lumbar region of the spine to assume a kyphosis, or posterior concavity, when seating occurs on an ordinary flat seat pan in the absence of adequate back support. This kyphosis is in opposition to a normal lordosis (bending in the anterior direction) which is seen in a standard upright posture. If such kyphotic postures continue, the results may be increased pressures on the lumbar disks, with concomitant softening, posterior protrusion, and accompanying pain.

In order to deal with this problem, Kroemer and Robinette propose a number of solutions. (It should be emphasized that, throughout this discussion, we will concentrate only on a subset of controversial design recommendations. Many other considerations have general support and will not be considered here.) Prominent among these is a call for a specific physical support in the lumbar region of the back. This lumbar support pad would act to gently force the spine into a lordotic posture when the sitter leaned against it. This pad was recommended either alone, or as a protuberance on a full-size back rest. In addition, seat pan height was to be adjusted such that thigh-leg angles were to approximate 90 degrees, and that trunk-thigh angles were to be equal to, or greater than, 90 degrees. Finally, this article suggests the possibility of adjustable worksurface heights so that the elbow can be located at the keyboard height allowing forearms to be parallel to the floor.

Backwards leaning option. With the advent of the video display terminal into the office environment, recommendations such as those of Kroemer and Robinette (ref. 8) were adapted to the new workplace in a fairly straightforward fashion appearing in references such as Cakir, Hart, & Stewart (ref. 7) and Galitz (ref. 9). However, the generality of these recommendations started to be questioned by Grandjean and his colleagues.

In an earlier study by Hunting, Laubli, and Grandjean (ref. 10) the authors had obtained data in a field study which violated conventional wisdom in that VDT operators whose keyboard heights were higher than optimal (greater than 84 cm) reported fewer muscular complaints than those with lower keyboards. A recent field study by Grandjean, Hunting, and Piderman (ref. 4) attempted to examine the problem in more detail. In this study, 68 subjects in 4 companies utilized adjustable ergonomic VDT workstations (chairs and terminal stands) over a five-day period. The results were quite clear. The subjects did not adopt a standard upright posture which, according to all recommendations, should have been optimum for them. Instead, they tended to tilt their adjustable back rests towards the rear, lean back, and raise their forearms upwards. Average trunk inclinations were seen to be between 100 to 110 degrees, with average elbow angle at 99 degrees. Hence, what we seem to observe in this situation is standard posture rotated backwards about 15 degrees. This rotation was, of course, greatly facilitated by the availability of adjustable chairs with full back rests, terminal stands which allowed keyboard heights to be raised to correspond to the raised forearm heights, and an arm rest in front of the keyboard for further support.

In their conclusions, Grandjean and his colleagues take issue with the fundamental postural premises reported by Kroemer and Robinette (ref. 8). It is worth quoting this material directly:

> It is a fact that some orthopaedists recommend an upright posture with a slight lordose of the lower part of the spine. On the other side, a Swedish group of orthopaedists (Nachemson (ref. 10a) and Andersson (ref. 11)) measured the pressure inside intervertebral discs as well as the electric activity of the back muscles in relation to different sitting postures. When the backrest angle of the seat was opened from

> 90 degrees to 110 degrees, they recorded on their subjects an important decrease of the intervertebral disc pressure and of the electromyographic activity of the back.
>
> Similar results were observed by Yamaguchi and colleagues (ref. 11a) who also advise an angle between seat and backrest of 115 to 120 degrees as the best condition for relaxation of the spine. All of these results indicate that resting the back against a sloping backrest transfers a proportion of the weight of the upper part of the body to the backrest and reduces noticeably the physical load on the intervertebral discs as well as the static strain of the back muscles. (ref. 4, p. 32)

The anatomical references cited by Grandjean et al. have not yet become available to this author, so that the postural status of the lower spine while in the backwards leaning orientation has not been ascertained. Obviously, these references must be carefully scrutinized in view of their obvious contradictory indications to the kinds of anatomical considerations discussed in Kroemer and Robinette (ref. 8). However, one further point must be raised. The actual chair utilized in the Grandjean et al. (ref. 4) study is described as a Giroflex. In their discussion of the work of Grandjean and his colleagues, Kantowitz and Sorkin (ref. 12, p. 480) display a picture of a Giroflex chair, the design of which is attributed to Grandjean. The chair clearly possesses a lumbar support pad as part of a high back-rest configuration. The question is whether the lumbar support pad designed for an upright posture is still effective, or possibly counterproductive, when the trunk is tilted backwards 15 degrees.

Forward tilt option. The arguments of Grandjean and his colleagues are in nearly direct opposition to the forward tilt proposal of Mandal (refs. 13, 14). Mandal noted that much of the daily work (writing, typing, reading) done in a seated position at a desk requires the individual to bend forward, typically in order to bring his/her work materials into comfortable visual range. Many of his initial observations were of school children, who, he observed, in order to minimize the hip flexion associated with such bending, tended to move to the forward portion of their seats. However, while this posture becomes visually advantageous, the front edge of the chair, acting as a fulcrum, digs into sciatic nerve and femoral vein. In addition, such a posture reverses the normally desirable spinal lordosis into an opposite kyphosis, with the ill side effects as described earlier.

The solution proposed by Mandal is to rotate the seat pan angle forward by 15 degrees. The result effect, theoretically, is that the pelvis tilts forward, and a lumbar lordosis is automatically restored. In addition, hip angle is kept within reasonable bounds (less than 90 degrees.) Under these conditions, the back rest is unnecessary, except to lean back during breaks. Mandal indicates that this posture may result in a tendency by the individual to feel that she/he is slipping forward, but argues that this may be eliminated by the proper use of fabrics. (In fact, as Kroemer and Robinette (ref. 8) indicate, similar proposals had been around since 1884, but objected to on the basis of the slipping problem.)

An investigation is described in which 10 individuals worked at 5 different workstations. These included typing at a normal (5 degree backwards slope) chair, reading and writing at a normal chair with the full extent of the seat pan utilized, reading and writing at a normal chair with the forward half of the seat pan utilized, reading and writing at a Mandal 15 degree forward tilt chair, and, finally, the previous condition with the addition of a slanted work-surface in order to minimize the residual forward bending. (In this connection, it must be indicated that the rationale for the traditional backwards slope is to force the individual back against the backrest.) Mandal measured elongation of the dorsal spinal muscle and pressure across the front, middle, and back of the seat pan. He found that elongation was nil in the typing posture; this served as a baseline since the operator sat with the spine in an erect posture. However, elongation was maximum (41 cm) in the second condition (reading/writing in normal chair), successively decreasing down to 0.7 cm in the fifth condition. With regard to pressure distribution, the first three conditions yielded a strong bias towards the first two-thirds of the seat pan; however, pressures were equalized under the forward tilt conditions. Accordingly, the forward tilt provides, it is argued, the optimal design solution.

Bendix and Biering-Sorensen (ref. 15) conducted a laboratory evaluation of forward tilt chairs. Ten subjects were seated an hour a day at each of four backless height and tilt-adjustable seats. Forward tilt values of seats were 0, 5, 10, and 15 degrees. During the experimental period, subjects sat with their elbows on a height-adjustable table and read a horizontally-placed text. Six times throughout each session, a series of angle and length measurements (statometric) were taken of various anatomical landmarks of the spine. The authors indicate a systematic increase in lumbar angle with increasing chair tilt (4 degrees over the 15 degree tilt range). Lumbar angle is defined as the angle between two vectors; the sacral-lumbar vector, and the lumbar-thoracic vectors. Vector endpoints were, as indicated, defined in terms of spinal processes. This increase is taken to indicate a systematic progression towards lumbar lordosis with increasing tilt angle, and would, therefore, seem to support the Mandal argument.

However, this finding was qualified by a supplemental investigation using a 15-degree tilted seat in which the relationship between hip angle and lumbar adaptation was evaluated. Hip angle was first estimated while subjects sat on a horizontal seat by measuring the patellar-anterior iliac distance. The seat was then tilted 15 degrees, and the subject's trunk tilted forward until the patellar-iliac distance was equal to that obtained in the horizontal condition. Measurement of lumbar angle in both conditions yielded an increase of 13 degrees from horizontal to 15 degree forward tilt. Comparing this with the 4 degrees obtained earlier, when the subjects could freely choose their posture, led the authors to conclude that, in the latter (normal) case, the majority (66%) of the bodily adaptation to the tilt took place as result of a backward pelvic rotation/hip extension, with only the remaining 33% occurring in the spine.

It was of interest that subjective estimates given by the subjects indicated that the 5-degree tilt seat was seen as most comfortable. This fact, together with the observation that the sacral-lumbar and sacral-thoracic vectors were more vertical in this condition than any other, led the authors to suggest that this smaller degree of forward tilt be the focus of additional research attention.

Articulated rear wedge. As discussed in Kroemer and Robinette (ref. 8), the so-called Schneider-wedge has been proposed as a third alternative design for enhancing postural effectiveness. As originally conceptualized, this involves an elevation of the rear portion of the seat to about 30 degrees, but with front part of the seat horizontal. This rear wedge has the effect of tilting the pelvis forward, as in the Mandal chair, but the horizontal front prevents the feeling of sliding forward. Kroemer and Robinette report evidence casting doubt on the effectiveness of the original Schneider design. However, a modification is now available with a less drastic wedge articulated with a full back rest which, according to sales literature available from the manufacturer (Martin Stoll), meets the objections described in Kroemer and Robinette (ref. 8).

Dynamic adjustability. Several new chair models are now available employing mechanisms which dynamically respond to changes in operator posture (e.g., Steelcase Concentrix, Cyborg). The author has discovered no literature evaluating these designs, but they will surely have an impact on the user community, and must be included in future ergonomic assessments and research efforts.

## ASSESSMENT OF ISSUES

It may be of some interest to trace the course of Kroemer's recommendations regarding chairs over a 15-year period. In Kroemer and Robinette (ref. 8), we see what appears to be the basis for the current "standard" recommendation, embodying an upright posture, lumbar support as an aid to lordosis, and a fundamentally horizontal seat pan. The full back rest is mentioned but only as an option. In Kroemer (ref. 16), a full back with lumbar support pad incorporated is now the recommendation of choice, with adjustability of the support region being stressed. In this paper, we also now find an emphasis on the desirability of allowing the seat pan angle to be adjustable, presumably to allow the Mandal option for those who prefer it. A shorter article in the same year by Kroemer and Price (ref. 17) presents the same argument, but, interestingly, omits any mention of lumbar support. Finally, Kroemer (ref. 18) presents a brief review of orthopedic theories of posture, then argues that such theories have ignored what seated people actually do (cf. the field study of Grandjean et al. (ref. 4)). He stresses the need for flexibility and adjustability in order that the seated individual may change posture easily.

Impressions and issues. Given the above progression of recommendations, one is left with the impression that the biomechanics of seating are not at all well understood. This lack of understanding seems to focus around two key questions: (a) what are the relative relationships between lumbar spine and pelvis which are least stressful during prolonged seating; and (b) what are the essential design features which will optimize these relationships? Given, however, the fact that information regarding biomechanical issues reviewed herein is all obtained from secondary sources, we are left with three possibilities:

(1) There is a fundamental lack of understanding of seated posture at the level of musculoskeletal functioning.

(2) The basic understanding of such functions exist among specialists in the field but has not been well understood by some ergonomists, hence, the presence of conflicting recommendations.

(3) These functions are, in fact, well understood by ergonomists to the extent that the conflicting recommendations are alternative paths to the same biomechanical end.

The purpose of writing this paper has, in fact, been to stimulate discussion towards the resolution of these alternatives. However, regardless of which of these alternatives is true, there are certain additional issues which must be considered in the practical context of seating recommendations, particularly in view of the costs involved in large scale replacement of office chairs.

(1) The criterion problem. What is the appropriate end point for design? Shall it be biomechanical (e.g., interdisc pressure changes), muscular (e.g., EMG changes), behavioral (e.g., performance efficiency), or psychological (e.g., subjective ratings of comfort)? These criteria may, of course converge, but what if they do not?

(2) The time course problem. What is the appropriate time course for measurement? It is reasonable to assume a certain period of adaptation (hours? days?) to a new chair after which a relatively steady-state biomechanical status exists? Has this factor been taken into account in current research, or

are indicated effects merely potential artifacts of initial unfamiliarity with new chairs? In particular, are there differential adaptational effects occurring in those experiments where the same subjects rapidly move between different chair types?

(3) The recommendation problem. What do we tell the user/manufacturer who can't wait for the research results/theoretical disputes to be resolved?

There is, of course, a considerable body of (relatively) noncontroversial information available for incorporation into design principles. However, with regard to the issues discussed here, the most reasonable approach is that of Kroemer (ref. 17), which emphasizes flexibility and adjustability. It does not seem that one can go too far off course in utilizing chairs with such flexibility, particularly compared with those which have been available in the past.

REFERENCES


1 E. N. Corlett, Pain and posture in E. N. Corlett and J. Richardson (Eds.) Stress, work design, and productivity, Wiley, London, 1981, pp. 27-42.

2 M. Dainoff, Occupational stress factors in visual display terminal (VDT) operation: A review of emperical research. Behavior and Information Technology, 1, (1982) 141-176.

3 S. Sauter, M. S. Gottlieb, K. C. Jones, V. Dodson and K. M. Rohrer, Job and health implications of VDT use: initial results of the Wisconsin NIOSH study, Communications of the Association for Computing Machinery 1983, in press.

4 E. Grandjean, W. Hunting and M. Piderman, VDT workstation design: preferred settings and their effects, Human Factors, 25, (1983) 161-175.

5 M. J. Dainoff, L. Fraser and B. J. Taylor, Visual, musculoskeletal, and performance differences between good and poor VDT workstations: preliminary findings, Proceedings of the Human Factors Society Annual Meeting, Seattle, 1982.

6 A. Cakir, D. J. Hart and T. F. M. Stewart, Video display terminals, Wiley, New York, 1976.

7 T. J. Springer, VDT workstations: a comparative evaluation of alternatives, Applied Ergonomics, 13.3, (1982), 211-212.

8 K. H. E. Kroemer and J. Robinette, Ergonomics in the design of office furniture, Industrial Medicine, 38, (1969), 115-125.

9 W. O. Galitz, Human factors in office automation, Life Office Management Association, Atlanta, 1980.

10 W. Hunting, T. Laubli and E. Grandjean, Constrained postures of VDU operators, in E. Grandjean and E. Vigliani (Eds.), Ergonomic aspects of visual display terminals, Taylor & Francis, London, 1980.

10a A. Nachemson and G. Elfstrom, Intravital dynamic pressure measurements in lumbar discs, Scandinavian Journal of Rehabilitation Medicine: Suppl. 1. (1970).

11 B. J. G. Andersson and R. Ortengren, Lumbar disc pressure and myoelectric back muscle activity, Scandinavian Journal of Rehabilitation Medicine, 3, (1974), 115-121.

11a Y. Yamaguchi, F. Umezawa and Y. Ishinada, Sitting posture: an electromyographic study on healthy and nostalgic people, Journal of the Japanese Orthopedic Assoc., 46, (1972), 51-56.

12 B. H. Kantowitz and R. D. Sorkin, Human Factors, Wiley, New York, 1983.

13 A. C. Mandal, The seated man (Homo Sedans), The seated work position. Theory and practice, Applied Ergonomics, 12.1, (1981), 19-26.

14 A. Mandal, Work-chair with tilting seat, Ergonomics, 19, (1976) 157-164.
15 T. Bendix and F. Biering-Sorensen, Posture of the trunk when sitting on forward inclining seats, Scandinavian Journal of Rehabilitation Medicine, in press, 1983.
16 K. H. E. Kroemer, Ergonomic recommendations for design of VDT workstations, Virginia Polytechnic Institute and State University, 1982.
17 K. H. E. Kroemer and D. L. Price, Ergonomics in the office: comfortable workstations allow maximum productivity, Industrial Engineering, 14, 1982, 24-32.
18 K. H. E. Kroemer, Ergonomics of VDU workplaces, Virginia Polytechnic Institute and State University, 1983.

# HUMAN FACTORS EPIDEMIOLOGY: AN INTEGRATED APPROACH TO THE STUDY OF HEALTH ISSUES IN OFFICE WORK

BENJAMIN C. AMICK, III DAVID D. CELENTANO, Sc.D.
Department of Behavioral Sciences
The Johns Hopkins University
School of Hygiene and Public Health
615 N. Wolfe St.
Baltimore, MD 21205

As office work becomes the dominant structural form of work in postindustrial society, greater emphasis will be placed on understanding the epidemiology of illness and disease among office workers. Rapid technological change from human-mediated work to computer-mediated work leads to changes in how workers adapt to their particular work environment (ref. 1,2). Concommitant with these technological shifts are societal transformations from industrial to postindustrial society (ref. 3), reflecting changes from a society based on a manufacturing economy to a society based on a service economy (ref. 4). Organizational policies, adjusting to these sociocultural transformations, place limits on how workers may carry out their work (ref. 5). Recently, Shaiken (ref. 6), in examining these issues and their attendent complexities, poses as the central research question on office automation, "...how do we evaluate the extraordinary transformations taking place around us?" Each transformation separately and together produces changes in the nature of work, especially in the office environment, demanding a new conceptual and methodological agenda for research into the health consequences of workers adapting to an automated office environment. The concept of a human factors epidemiology as proposed in this chapter should be considered in light of these transformations and the need for new methods to examine their influence.

Just as the shift from infectious to chronic disease epidemiology has required scientists to develop new approaches for understanding the origins and patterns of these decreases, so too will we need to develop new ways of understanding the origins and patterns of illnesses and diseases produced by changes in technology from a human-mediated work process to a computer-mediated work process. Social epidemiology and human factors research have been two independent methodologies and approaches examining the impact of technological change on worker health and safety which need to be integrated for assessing the health implications of new technology.

Focusing on the individual's health, social epidemiology relates the occupation, job category or characteristics of the requirements of specific occupations to worker illness and disease experiences most frequently employing a stress model. Alternatively, focusing on the work environment, human factors research describes the worker-environment interface, trying to optimize satisfaction, motivation and productivity. It is apparent that each of these approaches separately is inadequate to satisfactorly define and account for the health consequences of office work. Together, these approaches can be complementary and could provide a more thorough understanding of the origins and patterns of illness and disease in the changing office environment. Human factors epidemiology as a research perspective is not intended to be a unitary approach; rather, it is conceptualized as an integration of several research approaches with the eventual goal of more clearly understanding the potential contribution of aspects of the office environment to the quality of life, morbidity and mortality of workers.

Human factors research and social epidemiology have unique approaches which are inherently complementary. Human factors research typically employs a systems perspective in examining the worker-environment interface. This paradigm considers the individual and the work environment as a totality, a perspective not typical of social epidemiology. Epidemiology does not typically break down an occupation or job category into task components to evaluate their impact or to examine various physical characteristics of the work environment (e.g., workstation design) as contributing factors in subsequent worker illness and disease. However, such a decomposition is essential for understanding the role that technology will have in affecting a worker's health. Epidemiology has provided a coherent research approach for ascertaining worker health problems, integrating laboratory and field research. This epidemiologic method is founded on models of disease causation which relate characteristics of the environment to the development of disease, generating statements about an individual's risk of developing certain illness or disease outcomes.

The goal of this chapter is to introduce a human factors epidemiology perspective in general and to examine this approach as it applies to office automation. Following a discussion of models of disease causation, two components of human factors research, task variability and workstation design, are introduced and integrated into the epidemiological approach. Finally, a general model of technological change is presented, providing a framework for using a human factors epidemiological perspective.

## DEFINING HUMAN FACTORS EPIDEMIOLOGY: BRIDGING FIELDS

Human factors epidemiology is more than just the bridging of the fields of epidemiology and human factors . A <u>human</u> factors epidemiological approach implies that the object of study and knowledge is the human being, the individual worker. Explanations can be derived from biological, psychological or social process, but prevention is for the sake of humans.

### Epidemiology

Epidemiology is defined as the study of the patterns of health and disease occurrence in human populations and the factors that influence and predict these patterns (ref. 7). One central concept underlying epidemiological research is the natural history of disease. The development of disease is conceived of as a continuous process from individual exposure to a particular

MODEL OF THE NATURAL HISTORY OF DISEASE

| Exposure to agent | Initiation of the Etiologic Process | Pathological changes | Symptom Development | Clinical Recognition of symptoms and diagnosis of illness or disease | Outcome of Illness or disease process |
|---|---|---|---|---|---|
| --------- | Subclinical manifestations of illness or disease | | ----------- | Clinical manifestations of illness or disease | |

Figure 1: Exposure to an agent leads to initiation of the etiologic process which eventually results in pathological changes in the human body. Eventually, these changes manifest themselves as symptoms and signs concordant with clinical recognition of illness or disease. These result in some morbid or mortal outcome.

agent leading to an illness or disease outcome, as represented in Figure 1. "A causal relationship would be recognized to exist whenever evidence indicates that the factors form part of the complex of circumstances that increases the probability of the occurrence of disease and that a diminution of one or more of these factors decreases the frequency of that disease." (ref. 8) Such a multicausal view presupposes that a set of factors is associated with each disease. For example, in identifying the natural history of cardiovascular disease, high blood pressure, smoking and various lipoprotein fractions have been shown to be involved in the atherosclerotic process, yet biological factors alone are inadequate to explain the development of disease. Social epidemiologists (ref. 9) have argued that there are patterns of behavior which place people in contact with or at elevated risk of exposure to pathogenic agents (refs. 10, 11).

The differences between a multicausal view and one including human behavior in disease etiology result from differing conceptualizations of the natural history of disease (ref. 12). For example, Lilienfeld (ref. 13), in pointing out the need for a multicausal view of the origins of cancer, suggests: "Epidemiologic studies have strongly suggested that a vast majority (80 to 90%) of cancers are caused by radiation, chemical, and biologic agents, the remainder result from endogenous or genetic factors." (ref. 13) Lilienfeld distinguishes between the macro-environment, which could be considered the office and the people in it, and the micro-environment where an individual is exposed to particular agents such as asbestos from ceiling tiles which may initiate the etiologic process. Cox and McKay (ref. 14) extend this model to include psychosocial stressors and point out: "It seems increasingly clear however, that there are large behavioral components which govern exposure to potential carcinogens and there is growing interest in the extent to which social and psychological demands may be associated with these agents or may operate as contributory factors in their own right." (ref. 14) Much in the same way that continual exposure to a carcinogen may determine the development of cancer, daily demands in the office may initiate the etiologic process or cause changes in behavior (life-style) such that exposures are altered. This psychosomatic model of disease causation (ref. 12) extends the model in Figure 1 by placing psychosocial factors with environmental and biological agents which initiate the etiologic process. To be integrated into the natural history of disease the biological consequences of psychosocial processes must be addressed. The main areas of this active research are the so-called stress-related diseases (cardiovascular, gastrointestinal and mental). There is a growing literature on the relationship between psychosocial stressors and neurohormonal and immunological processes (ref. 44, 15). The work of Sterling and Eyer (ref. 16), in discussing the biological basis of stress-related mortality, shows how environmental demands elicit numerous biochemical responses and through a complex series of pathways involving many organ systems initiates pathological processes.

The concept of neurohormonal arousal is central to this model. Cardiovascular pathology has been the focus of much of this research (refs. 15, 16, 17), with recent research examining the role of psychoneuroendocrine responses in cancer pathology (ref. 14). Arousal is thus an intermediary between psychosocial events and symptoms, illness and disease. One characteristic of those diseases within a psychosomatic model is long duration associated with chronic arousal. The process of neurohormonal stimulation is then one of a chronic cumulative nature occuring over time. Patterns of behavior eliciting chronic arousal associated with daily work routines and office interactions then become potential agents in the natural history of disease. Research has provided initial support for this process, but problems remain. Critics claim, "they have not shown that the biochemical and physiological changes which are a consequence of stress are of sufficient magnitude to cause the range of diseases involved." (ref. 12) Many of the correlations between stressful experience and measures of neurohormonal arousal are often low and sometimes contradictory (ref. 11). Others have cited the protective effects of catecholamine release during physical activity (ref. 22). Despite these limitations the psychosomatic model provides a plausible link between stress and biological processes. Arguing from this perspective, daily stresses of office work such as lack of supervisor support or low status, stimulate chronic arousal which over time initiates pathological processes leading to the development of chronic diseases.

In those studies investigating etiologic factors only rarely are multiple factors in the etiologic process conceptually portrayed as interacting. The interaction of VDT radiation and stress in reproductive outcomes, or the role of stress in exacerbating an individual's chance of developing acute illnesses are neglected. Possible examples of the latter in the office are: (1) the interaction of stress and chemicals from carbonless copying paper in producing skin, eye or mucous membrane problems in office workers (ref. 19), and (2) the interaction of poor chair design and the stress of a continuous work process in lower back problems. It can be argued that the consequences of these short-term illnesses are chronic arousal, stressors for long-term illness. Supporting this is recent evidence demonstrating that the best predictor of mental distress is previous illness (ref. 20).

Cassel (ref 21) has proposed a general susceptibility model of disease causation which emphasizes the examination of multiple causes and multiple outcomes, in which susceptibility to disease is linked to decreased resistance to physiochemical and microbiological agents. An interaction occurs between psychosocial processes and these agents. Chronic arousal remains the intermediary between worker behavior and some pathological event. Extending this paradigm to the work environment, an interaction occurs between psychosocial events and physical and structural characteristics of the work environment. Given the introduction of new technologies into the office environment whose health hazards have not been comprehensively specified, a less specific model seems warranted.

The general susceptibility model of disease causation qualitatively differs from the psychosomatic model. It proposes multiple causal pathways in the natural history of disease to the initiation of pathology, each contingent on the physical and structural characteristics of the work environment. Rather than constructing psychosocial events which insult the body, eventually initiating some pathological process, psychosocial processes alter an individual's susceptibility by modifying the neurohormonal balance. Exposure to an agent becomes a broader system depicting the worker-environment interaction. Subsequent to workers carrying out their job (adapting to their work process) may be alterations in the neurohormomal balance. Potentially, risk factor constellations, reflecting these patterns of behavior, can be associated with multiple outcomes (ref. 22). To estimate the adverse health consequences of recent technological changes from human-mediated to computer-mediated work and to monitor the long-term impact of office automation, this model readily captures the array of potential consequences in the multiple-risk factor--multiple-health outcome perspective of the general susceptibility model of disease causation.

There are several advantages of this model over the more traditional psychosomatic model. First, stress becomes a term describing many psychosocial processes, thereby avoiding measurement confusion (ref. 23). Second, it does not imply a dose-response relationship (ref. 21). Finally, the responsibility for the prevention of adverse health states does not reside solely with workers or management, encouraging worker/management negotiation.

Several recent studies have provided initial confirmation of the general susceptibilty model of disease causation. Najman (ref. 12) asserts this model best explains the patterns of morbidity and mortality in the United States (ref. 12), and occupational mortality in Australia (ref. 24). The integration of psychosocial processes into the natural history of disease by social epidemiologists facilitates incorporation of the human factors approach to the worker-environment interaction. Recently, in a review of social factors in chronic disease, McQueen and Siegrist (ref. 17) cited several advantages and disadvantages to epidemiologic research. Two problems noted were: (1) the problem of the oversimplification of social variables; and, (2) the failure to consider multicausal relationships among both the causes and the outcomes. The latter problem is partly a consequence of a lack of a systems perspective and partly the model of disease causation implicitly used, while the former problem is an oversimplification of the worker-environment interaction.

The relationships of physical and structural characteristics of the work environment to psychosocial processes, and ultimately to disease needs to be expressed. The remainder of this section deals with an approach to this issue using human factors research.

### Human Factors

The use of the term 'ergonomics' in lieu of human factors has recently become popular in the U.S. Ergonomics, in its broadest definition, is the study of social, physical and psychological characteristics of man in his relation to work. Other chapters in this book discuss ergonomic issues in office automation. This systems perspective examining the complete worker-environment interaction has been discussed as applicable in the general susceptibility model of disease causation. This section will draw from human factors research the approach which elaborates the complexity of the worker-environment interaction.

McCormick (ref. 25) defines human factors engineering as, "...the application of information about human beings and their capabilities and limitations to the design of equipment which people use and to the environments within which people live and work." The goal is to optimize the interaction between workers and the work environment maximizing productivity, motivation and satisfaction. In office automation crucial areas of investigation are the tasks performed, workstation design and their interaction (refs. 26, 27).

Task Variability

A task is any activity the worker performs during the course of the day in the work process. To do these tasks the worker must use the available technologies in the work environment. Tasks performed on a regular basis are of primary interest. These patterned, daily behaviors reflect specific worker-environment interactions which can (potentially) initiate chronic arousal in individuals. It is important that the focus be on tasks performed on a routine basis rather than a single task performed rarely. This is not meant to deny the significance of single events, particularly in light of the meaning individuals attach to them. Work is conceived as labor over time during which tasks may be performed many times or once. (See Pearlin (ref. 28) for a similar discussion of the relationships between chronic stressful events and acute stressful events in the stress process.) A common assumption in epidemiological research is that people in the same job do the same tasks. Task variablility is a neglected area of study, along with the processual nature of work.

Tasks are characteristic of the work process, representing one form of worker-environment interaction. Task variability is structurally determined either organizationally or socially. An example of how organizational structure determines task variability is the work of telephone operators: operators who work on computer terminals have a defined number of customer calls they should take in one hour based on parameters used by a computer which forwards calls to the operator. By selecting this particular technology (computer and software) the organization has defined the nature of the task, as well as the task variability. Task variability describes the interplay between workers and various technologies in the office environment.

Two characteristics of task variability are important for a description of the worker environment interaction: (1) daily individual patterned task variability, and (2) task variability performing similar work with different technologies.

Table 1 : Percent of Health Problems Across Technologies And by the Type of Task

| | VDT Users | | Non VDT Users | |
|---|---|---|---|---|
| Health Complaint | Data Entry | Conversational | Typist | Other Office Workers |
| Sore Shoulders (ref. 32) | 60% | 30% | 25% | 10% |
| Visual Complaints (ref. 33) | 65% | 72% | 55% | 50% |

Research on the patterned daily tasks of office workers is limited (ref. 29), especially in relation to patterns of illness and disease (ref. 31). Table 1 summarizes the results of studies which have reported health complaints of VDT users; it is evident that people using a VDT do not necessarily perform the same tasks. Smith (ref. 26) has emphasized the need to examine computer-mediated work by the type of task involved. Instead of assuming that in an automated office secretaries working on VDT's are equally exposed, the amount of time spent working on a VDT and the types of tasks performed (daily task variability) should be investigated and related to patterns of illness and disease. Damron (ref. 32) found that secretaries spent 50% of their time at the typewriter even though the offices were automated.

Daily patterned task variability is also important in the way it interacts with psychosocial processes. There is a large literature showing that lack of task variability is negatively related to individual health (ref. 33). This research has focused primarily on traditional machine paced jobs, with little study of the office. Considering the example of the telephone operator mentioned earlier, the lack of task variability (continually answering calls forwarded by the computer) can be considered monotonous by the operator, with little opportunity for social interaction. If a worker is responsible for these tasks on a daily basis and has little control over when work is completed, the interaction of high job demands and limited control can potentially result in continued chronic arousal, leading to illness and disease (ref. 34).

As the technology used in work changes so does the nature of the task (ref. 1, 2). Two secretaries typing a letter, one on a typewriter and the other on a word processor, are performing different tasks requiring different physical, social and mental capacities. Each of these tasks requires the worker to interact with a different set of characteristics of the work environment. Task variability between technologies provides a unique example of how the task is intertwined with psychosocial processes.

With the introduction of computer-mediated work, the task of the worker has been altered by changing the way one knows about the object of the task: the work has become abstract. Workers get feedback about the task object only as symbols through the medium of the information system. The object of the task has disappeared 'behind the screen' (ref. 2), leading to worker alienation (ref. 35). A NIOSH study shows that worker involvement in work is lowest among clerical workers, supporting the assertion of change in the mental nature of the task. Alienation has been shown to be related to mental health (ref. 36). As offices become increasingly automated, one can expect more alienation, resulting in increasing mental health problems.

Technological changes may also modify the social environment. The computer terminal can become the worker's primary focus of interaction. People no longer work with people, they work with video display terminals. This may lead to problems of social isolation and a lack of support from others. Smith (ref. 37) has shown that clerical workers using VDT's report lower levels of co-worker support and staff cohesion compared to professionals and clerical workers who do not work on VDT's. Social epidemiologists have shown the negative health problems associated with a lack of support (ref. 10): problems range from cardiovascular mortality (ref. 38) to depression (ref. 39). However, House and Wells (ref. 14) found co-worker support did not significantly affect illness outcomes in a study of a New England chemical factory. They attributed this to the organization of the work which did not allow employees to interact in any meaningful way.

The relationship between the worker and the supervisor also changes with the advent of computer-mediated work. Supervisors gain direct access to the employee's work. Monitoring allows supervisors almost immediate feedback about employees work, having both positive (ref. 40) and negative effects (ref. 2). One problem in the automated office is that the ability to monitor work without physically moving from a desk may reduce supervisor-employee interaction; Haynes (ref. 41) has shown this to be a strong predictor of cardiovascular mortality.

Finally, the changes in the nature of the task can alter the level of control an individual has over the process of work. Control can shift from the worker to the supervisor (ref. 1). Perceived supervisor control is greatest among clerical workers using VDT's, reflecting decreased worker control over tasks and daily patterned task variability. Karasek's research on worker control shows that lack of control over the work process is related to both mental strain and cardiovascular mortality (ref. 34). Similar results are found elsewhere (ref. 42). It can be suggested that task variability between technologies may potentially affect a worker's health by reorganizing the social structure of the work environment.

Changes in the task dictate changes in the physical requirements of the worker. The physical requirements for working at the VDT differ from working at a typewriter. The design of the workstation limits how the worker can interact with the technology. Therefore, task variability is directly related to workstation design, another area of human factors research important for describing the worker-environment interaction.

Workstation Design

The physical characteristics of the work setting limit how the worker can interact with various technologies. Human factors researchers have spent more time describing the optimal physical characteristics of the office environment for the worker than any other area of the worker-environment interaction, especially for computer-mediated work processes (ref. 43). In office environments the design of the workstation is fundamental for understanding how the worker can use various technologies. Since several chapters discuss how the workstation places limits on the way workers perform various tasks, this subsection only highlights health consequences of workstation design and the interaction of workstation design with psychosocial processes to produce health consequences.

In any office there are certain common physical requirements for the worker to carry out their work. One general list may include a chair, a table and lighting. Although not complete, these components provide a useful constellation of workstation design factors for describing the worker-environment interaction and the adverse health consequences. Recent work in Sweden has implicated static work posture in musculo-skeletal disorders (ref. 33). Swedish researchers have suggested preventive reorganization of the work environment (ref. 33). Yet, one cannot adopt a specific set of parameters for a chair, a table and lighting for all office environments, as these depend on the task performed and the technology used. For example, as offices install automation technologies, the table and lighting requirements change because the nature of the task changes. The National Academy of Science (ref. 44) recently enumerated the appropriate workstation design features needed in computer-mediated work. In most offices where automation has occurred the workstation has not been redesigned. They only change the technology used to complete the task, leading to reports of health problems among office workers. Sauter (ref. 45) found in office workers using VDT's that chair and workstation configuration were predictors of musculo-skeletal disorders, while ambient lighting was predictive of visuo-ocular disorders. These disorders develop over time as workers use various technologies in their work environment. As workers adapt to a work process these problems may become part of everyday working. Workers, in an attempt to deal with these problems, may release certain neurohormones entering a state of arousal. If this arousal persists over extended periods of time, then chronic arousal may lower an individual's susceptibility to an array of illnesses and diseases. (Frankenhauser (ref. 42) has termed this adaptive process 'healthy maladjustment'.) Acute musculo-skeletal and visuo-ocular problems may be exacerbated, resulting in chronic conditions which no longer require the worker to be performing the task to be present.

Features of the design of the workstation also may interact with psychosocial characteristics of the work process. Interaction may be direct; if a secretary is typing continuously for 2.5 hours with little control over when she can take a break, then the configuration of the workstation and the typing demands coupled with little control may interact to produce musculo-skeletal or visuo-ocular problems. Sauter (ref. 45) found that significant predictors of musculo-skeletal problems were chair comfort, the workstation configuration and job demands. There has been no other research in this area. One contribution of the general susceptibility model would be to encourage this type of research.

Human Factors Epidemiology

The approach outlined here draws from two fields unique approaches for describing the work process. Social epidemiologists relate psychosocial processes to acute and chronic health outcomes. Human factors researchers typically relate task variability and the design of the workstation to worker productivity, satisfaction and acute health problems. Together, they provide a more comprehensive approach for assessing long-term health consequences of technological change. This integration results in a broader description of the work process, one more effective in providing understanding of the changes in the work environment when technologies change.

As technology changes within the office workers are exposed to various structural and physical characteristics of the environment as they perform their job. Each job is composed of several tasks which the worker performs routinely. To carry out these tasks the worker uses specific technologies available in the organization. The physical design of their workstation limits how they physically manage their work. Over time the worker learns the amount of control they have over task variability, as well as the demands of each task. They develop patterns of work and interaction with co-workers and supervisors contingent on the technology they are using, task variability, workstation design and job demands. This complex system is the work process each individual is involved in during their daily work routine. Introducing a new technology into the system necessitates changes in other parts of the system for workers to effectively complete tasks. One of the challenges of studying office automation will be to describe the various changes in the system, capturing the form and content of the worker-environment interaction.

To specify risk statements about particular relationships between components of the worker-environment interaction and illness or disease epidemiologic research is essential. By demonstrating that various factors in the worker-environment interaction vary with illnesses and diseases, statements of risks can then be made. Risk factor constellations which describe particular worker-environment interactions can be related to the different types of office environments.

Human factors epidemiology describes the complexity of the worker-environment interaction, allowing specific relationships to be established between the physical and structural work environment, worker behavior and health.

REFERENCES

1 H. Braverman, Labor and Monopoly Capital: The Degradation of Work in the Twentieth Century, Monthly Review Press, New York, 1974.
2 S. Zuboff, New Worlds of Computer-Mediated Work, Harvard Business Review, 60:5 (1982) 142-152.
3 D. Bell, The Coming of the Post-Industrial Society, Basic Books, New York, 1973.
4 H. Levin and R. Rumberger, The Educational Implications of High Technology, Stanford University Institute for Research on Educational Finance and Governance, Stanford, CA, 1981.
5 O. Ostberg, Office Computerization: Research Then and Now, in D. Marschall and J. Gregory (eds.), Office Automation: Jekyll or Hyde?, Working Women Education Fund, Cleveland, Ohio, 1982, 127-142.
6 H. Shaiken, Choices in the Development of Office Automation, in D. Marschall and J. Gregory (eds.), Office Automation: Jekyll or Hyde?, Working Women Education Fund, Cleveland, Ohio, 1982, 5-14.
7 D. Kleinbaum, L. Kupper and H. Morgenstern, Epidemiologic Research & Principles and Quantitative Methods, Lifetime Learning, Belmont, California, 1982.
8 A. Lilienfeld and D. Lilienfeld, Foundations of Epidemiology, Oxford University Press, New York, New York, 1980.
9 S. Graham and L. Reeder, Social Epidemiology of Chronic Diseases, in H. Freeman, S. Levine and L. Reeder (eds.), Handbook of Medical Sociology, Prentice-Hall, Englewood Cliffs, New Jersey, 1979.
10 J. House and J. Wells, Occupational Stress, Social Support and Health, in A. Mclean, A. Black and M. Colligan (eds.), Reducing Occupational Stress, U.S. Governmant Printing Office (No. 78-140), Washington, D.C., 1978, 8-29.
11 R. Rose, C. Jenkins and M. Hurst, Air Traffic Controller Health Change Study, U.S. Department of Transportation, Washington, D.C., 1978.
12 J. Najman, Theories of Disease Causation and the Concept of a General Susceptibility: A Review, Social Science & Medicine, 14A (1980) 231-237.

13 A. Lilienfeld, Some Aspects of Cancer Epidemiology, Biometrics (Supplement) 1982, 155-160.

14 T. Cox and C. MacKay, Psychosocial Factors and Psychophysiological Mechanisms in the Aetiology and Development of Cancers, Social Science & Medicine, 16 (1982) 381-396.

15 J. Henry, The Relation of Social to Biological Processes in Diseases, Social Science & Medicine, 16 (1982) 369-380.

16 P. Sterling and J. Eyer, Biological Basis of Stress-Related Mortality, Social Science & Medicine, 15E (1981) 3-42.

17 D. McQueen and J. Siegrist, Social Factors in the Etiology of Chronic Disease: An Overview, Social Science & Medicine, 16 (1982) 353-367.

18 R. Kvetnansky, et al., (eds.), Catecholamines and Stress: Recent Advances, Elsevier, New York, New York, 1980, pp 557-571.

19 T. Menee, G. Asnaes and N. Hjorth, Skin and Mucous Membrane Problems from "no carbon required" Paper, Contact Dermatitis, 7 (1981) 72-76.

20 A. Williams, J. Ware and C. Donald, A Model of Mental Health, Life Events and Social Supports Applicable to General Populations, Journal of Health and Social Behavior, 22 (1981) 324-336.

21 J. Cassel, The Contribution of the Social Environment to Host Resistance, American Journal of Epidemiology, 104:2 (1976) 107-123.

22 D. McQueen and D. Celentano, Social Factors in the Etiology of Multiple Outcomes: The Case of Blood Pressure and Alcohol Consumption Patterns, Social Science & Medicine, 16 (1982) 397-418.

23 J. Cassel, Appraisal and Implications for Theoretical Development, in S. Syme and L. Reeder (eds.), Social Stress and Cardiovascular Disease, Milbank Memorial Fund Quarterly, 45:2 (1967) 41-45.

24 J. Najman and A. Congalton, Australian Occupational Mortality, 1965-67: Cause Specific or General Susceptibility, Sociology of Health and Illness, 1:2 (1979) 158-176.

25 E. McCormick, Human Factors Engineering, McGraw-Hill, New York, 1976.

26 M. Smith, Health Issues in VDT Work, National Institute for Occupational Safety and Health, Cincinnati, Ohio, 1981.

27 M. Helander, The Automated Office: A Description and Some Human Factors Design Considerations, Virginia Polytechnic Institute, Blacksburg, VA, 1983.

28 L. Pearlin, M. Lieberman, E. Menaghan and J. Mullan, The Stress Process, Journal of Health and Social Behavior, 22 (1981) 337-356.

29 W. Hunting, T. Laubli and E. Grandjean, Postural and Visual Loads at VDT Workplaces, Ergonomics, 24 (1981) 917-931.

30 T. Laubli, W. Hunting and E. Grandjean, Visual Impairment in VDU Operators Related to Environmental Conditions, in E. Grandjean and E. Vigliani (eds.), Ergonomic Aspects of Visual Display Terminals, Taylor and Francis, London, England, 1980, 85-94.

31 B. Cohen, Proceedings of a Conference on Occupational Health Issues Affecting Secretarial and Clerical Personnal, July 21-24, National Institute for Occupational Safety and Health, Cincinnati, Ohio, 1981.

32 J. Damron, An Evaluation of Office Automation in the World Bank, Unpublished Document.

33 A. Kilbom, Occupational Disorders of the Musculo-skeletal System, Newsletter of the Swedish National Board of Occupational Safety and Health, 1 (1983) 6-7.

34 R. Karasek, D. Baker, F. Marxer, A. Ahlbom and T. Theorell, Job Decision Latitude, Job Demands, and Cardiovascular Disease: A Prospective Study of Swedish Men, American Journal of Public Health, 71:7 (1981) 694-705.

35 R. Blauner, Alienation and Freedom: The Factory Worker and His Industry, University of Chicago Press, Chicago, 1964.

36 A. Kornhauser, Mental Health of the Industrial Worker, Wiley, New York, 1965.

37 M. Smith, B. Cohen and L. Stammerjohn, An Investigation of Health Complaints and Job Stress in Video Display Operations, Human Factors 23:4 (1981) 387-400.

38 L. Berkman and S. Syme, Social Networks, Host Resistance and Mortality: A Nine-Year Follow-Up Study of Alameda County Residents, American Journal of Epidemiology, 109:2 (1979) 186-204.
39 G. Brown and T. Harris, Social Origins of Depression: A Study of Psychiatric Disorder in Women, Free Press, New York, 1978.
40 J. Hackman and G. Oldham, Work Redesign, Addison-Wesley, Reading, Mass, 1980.
41 S. Haynes and M. Feinleib, Women, Work and Coronary Heart Disease: Prospective Findings from the Framingham Heart Study, American Journal of Public Health, 70 (1980) 133-141.
42 M. Frankenhauser, Coping with Stress at Work, International Journal of Health Services, 11:4 (1981) 491-510.
43 E. Grandjean and E. Vigliani (eds.), Ergonomic Aspects of Visual Display Terminals, Taylor and Francis, London, England, 1980.
44 National Academy of Science, Video Displays, Work and Vision, National Academy Press, Washington, D.C., 1983.
45 S. Sauter, et al., Job and Health Implications of VDT Use: Initial Results of the Wisconsin-NIOSH Study, Communications of the ACM, 26:4 (1983) 284-294.

## SECTION 4
## Physiological and Psychological Effects of Office Work

# MENTAL HEALTH OF SECRETARIAL AND CLERICAL PERSONNEL

JACQUELYN H. HALL
Mental Health Education Branch
NIMH
5611 Fishers Lane, Room 15C17
Rockville, Maryland 21857

The systematic study of occupations is fascinating, and it is particularly intriguing to note how such studies reflect the values that society holds. For years behavioral and social scientists have studied the psychology of careers and work among physicians, engineers, lawyers, military officers, corporate executives, and others who hold positions that rank high on the career prestige hierarchy. Earlier studies were made almost exclusively of men, so that basic theories about the psychology of career choice and occupational development reflected a male model. As for health matters, in any library the shelf that deals with occupational health holds a number of books on such subjects as "executive stress," "manager's stress," and other similar titles that imply that people who make the policy decisions in organizations suffer the most job-related stress. Also with new opportunities opening, considerable attention has been paid to the relatively new cadre of highly successful women who are breaking into executive suites.

Only recently have social and behavioral scientists delved into studies of two occupational categories that comprise by far the largest number of people, mostly women, and do not command high prestige ratings: the occupations of homemaking and office work. Indeed, until the women's movement began to flourish, most people referred to housewives as "nonworking," and even a recent Supreme Court decision about military pensions underscores the notion that homemaking and childrearing have no economic value.

But office workers, not homemakers, are the subject of this report. Clerical work now represents nearly one in five members of the U. S. work force, and accounts for more than one in three women who work. In short, most secretarial and clerical workers are women who are both office workers and homemakers, and research has shown that they do most of the housework and child care in their respective homes, even if they have male partners and work full time.

Interestingly, scientists now are finding that some of the most serious symptoms of stress appear among workers in clerical and secretarial roles.

The kinds of stress endured by executives have been studied extensively. The kinds of stress endured by office workers have only begun to be studied systematically, and few studies have included detailed measures of mental health among the variables under investigation.

## CORONARY HEART DISEASE

The famous Framingham study of heart disease (ref. 1) includes an 8-year study of the relationship of employment status, employment behaviors, and incidence of heart disease in working women, housewives, and men. In this study, it was found that when all women were considered together, employment status did not affect the risk of developing heart disease. When occupational groups were studied separately, however--delineating white-collar, blue-collar, and clerical workers--the results were stunning: among women, clerical workers were almost twice as likely to develop heart disease as managerial, professional, service, or blue-collar workers. Furthermore, the rate of heart disease among women clerical workers was about twice as high as the rate among housewives. This is an interesting contrast to heart disease among men, which is greater among white-collar workers. Another interesting note is that single working women exhibit the lowest rate of heart disease of all groups in the study. The women at highest risk for heart disease were women in clerical jobs, with little or no job mobility, with nonsupportive bosses, for whom the women felt a lot of suppressed hostility that they did not discuss. Those at highest risk were married to blue-collar husbands and had family responsibilities--in fact, the greater the number of children a woman had, the greater her risk for heart disease.

## MENTAL HEALTH AND GENERAL WELL-BEING

Depression is a problem that is much more prevalent among women than among men. Women employed in low-level jobs, especially those who are heads of households with small children, are at high risk for depression. Still, it is the unemployed women who show more symptoms of stress and more mental health problems than working women.

Other approaches to studying women and work show that depressed women who are employed outside of the home function better than depressed housewives who do not have paid employment. This research even suggests that outside employment may have a protective effect for some women's emotional well-being.

## ALCOHOL AND DRUG PROBLEMS

When drinking or other drug abuse interferes with work performance, office workers are likely to be considered expendable and to be released from their jobs. Therefore, they are not likely to be confronted about their problems and referred for treatment. Management-level workers, however, are more likely to be viewed as "worth the company's investment" in treatment or rehabilitation for such conditions.

The greater stigma for women alcoholics and drug abusers makes the issue even more complicated. When women have alcohol problems, they frequently are not diagnosed accurately by physicians, and physicians compound the alcohol problem with prescription drugs. When a woman is hooked on prescription or illicit drugs, by the time she enters treatment she may have a host of physiological problems, may be separated from family and friends, may have been arrested, and her self-esteem is about as low as it can go. She may fear that a social service agency will take away her children. In short, the double standard of stigma associated with addiction has closeted, and may have killed, many women office workers who are heavy drinkers or drug users (ref. 2).

## COMPLICATING EFFECTS OF MINORITY STATUS

While new findings about stress among office workers are mounting--and are being met with resistant disbelief by many managers--there still is almost no research on job stress experienced by minority workers.

Stereotyped expectations of racial and ethnic groups, limited command of English language, prevalence of hypertension among the Black population, and race discrimination are examples of complicating factors that have not been considered in basic studies of job stress. In fact, sometimes these factors have been used to omit minority workers from sample populations under study. Such factors now lead a number of researchers to speculate that they will find more harmful health effects from job stress among minority workers than have been found among other groups.

## CHRONIC STRESS

There are a number of research studies now suggesting that chronic stress has a profound influence over one's health status. This is in contrast to some classic research exemplified by the work of Holmes and Rahe, (ref.3) which emphasizes the results of stress caused by major life events. We see now that research in several parts of the country indicates that chronic stressful conditions at work or at home may be more salient predictors of mental illness and emotional dysfunction than traumatic and special life events. There is also some indication that this sort of finding may be more pertinent to women than it is to men.

Another kind of stress that appears to be more salient for women is "contagious" stress--that is, women experience stress because of other people's problems more often than men do.

It is well known that several stresses together have a much greater effect than can be estimated in simple additive terms. Physical stress, nervous system strain, intellectual fatigue, and psychological pressure cannot be neatly separated from each other. When they are combined, the stress is greater than the sum of its parts. That is, the combined stress creates a health risk that is greater than might be assumed if one simply added the risks together of any set of single elements. And chronic, continuous stress poses a severe danger to any person's health and well-being.

## SEXUAL HARASSMENT

Sexual harassment is finally being recognized as a significant barrier to economic advancement and a potent source of job stress. Studies (ref. 4) show that victims of such harassment typically are women in clerical and service roles; harassers typically are older men who have the power to hire and fire the women. In cases of sexual harassment, women report that it seriously undermines their well-being. They report a variety of symptoms, including:

Emotional stress--nervousness, fear, anger, sleeplessness, embarrassment

Interference with job performance--distraction, avoidance, loss of motivation

Physical stress--headaches, nausea, weight changes

Need for professional therapeutic help

In short, sexual harassment is a hidden occupational hazard that can have devastating mental health effects.

## SOURCES OF JOB STRESS

There are several items that are generally agreed upon as key sources of job stress for clerical and secretarial workers:

### Heavy workload

There is often simply too much work to do in too little time. Secretarial and clerical personnel commonly find themselves in situations where the staff is insufficient to meet the unreasonably high demands for productivity. The quality of workload is also important. An especially bad combination is having too much work that holds little interest or challenge.

### Conflicting instructions

Office workers are often subjected to conflicting demands from a number of bosses.

### Too little work

Another kind of stress to be taken seriously is the stress that occurs when there is too little work to be done.

Long working hours

Some studies among men show that men who work longer hours are more subject to coronary heart disease than men who work no more than 40 hours a week. Complicating this, people who are in secretarial or clerical positions often find that they are called upon for frequent and unexpected overtime assignments. Exceptionally long working hours for women result in extra stress because of their dual role as homemakers and wage earners.

Rapid pacing and machine-paced work

Office work traditionally involves meeting deadlines, making fast and frequent decisions, and carrying out quick transactions. Machine monitoring of productivity requirements makes this aspect of work life excessively demanding and extremely dehumanizing.

Monotony

Repetitive work is stressful; doing the same stereotyped motion all day causes muscular fatigue and job dissatisfaction. Also performing the same set of routines without the chance to learn new skills can lead to boredom, frustration, and lowered self-esteem.

Dealing with clients or the general public

For people who work in service institutions there are special pressures, particularly for those office workers who deal with the public by telephone or in person. Those who are in constant contact with clients frequently have to handle complaints and must be sensitive to personal and political situations; yet these clerical and secretarial personnel rarely have a part in making decisions that would solve any problems.

Discrimination

As an institutional stressor, discrimination may come through channels of sex, race, age, handicap, or combinations of these. Such discrimination is a chronic stress for many office workers.

Lack of recognition and respect

The lack of recognition and respect for valuable work is frequently reflected in low pay and unfair treatment.

Supervision problems

Nonsupportive bosses have been found to be correlated with coronary heart disease and other kinds of stress-related diseases among office workers.

Few opportunities for self-expression

Secretaries and clerks commonly feel that they cannot express anger, question accepted procedures, or contribute to decision-making and management.

Lack of job security

This is particularly important in an age where rapid introduction of new technology brings on rapid shifts in the number and character of clerical personnel needed.

Lack of privacy

Most office workers have desk areas that are in open space so they are in constant view of colleagues and supervisors. The lack of privacy is exacerbated when office workers have no access to a place where they can make or receive a telephone call privately, carry on a conversation, or relax quietly on a break.

No input into decision-making

Office workers are rarely asked for their views on policy and procedural matters or equipment installations--even those that have a substantial influence on their jobs. Furthermore, if they offer ideas, too often they simply are ignored--as if they had never spoken.

Isolation

Office workers frequently have little or no information about the overall workings of their own unit. They receive information about what has been decided, but all too often never why a specific decision was made.

Power relationships

In most work organizations, a hierarchical structure underlies the power relationship that exists, and within those power relationships some consistent patterns emerge:

(i) The power holders. Typically, the power holders define the subordinates as inferior in some kind of way; then the power holders treat them badly--acting out the assumptions of inferiority.

The power holders assign subservient roles and functions to the subordinates; these typically are things they do not want to do themselves. Power holders guard the highly valued functions for themselves--with an indication that the subordinates are really unable to perform those tasks.

They encourage the subordinates to develop characteristics that are pleasing to the powerful people. Most often these characteristics form a familiar cluster including submissiveness, passivity, dependency, lack of initiative, and limited ability to act or think or make decisions. If subordinates adopt these characteristics, they are considered to be well adjusted, good employees. Subordinates who show other characteristics--like intelligence, initiative, or assertiveness--are defined at least as unusual, if not downright abnormal, and organizational arrangements do not accommodate their competence.

Power holders usually limit the subordinates' freedom for experience, expression, and action; often the limits are described magnanimously as protections for the subordinates.

Power holders' opinions are not challenged, and they frequently receive flattery. Over time this leads them to the belief that their opinions are superior. Power holders consider themselves and their value system as the model for "normal."

Power holders are usually convinced that the way things are is right and good, not only for them but especially for the subordinates. Since they define organizational structures, philosophies, and policies, all those aspects of the workplace confirm their view.

(ii) The subordinates. Now think about the subordinate people: they must concentrate on basic survival. While they are often treated badly by the powerful people, they avoid a direct and honest reaction to the destructive treatment. They avoid open, self-initiated actions in their own interests; they resort to disguised and indirect ways of acting. They are labeled troublemakers if they question a set of rules and show free expression and action.

Historically some subordinate individuals didn't conform to the power-holders' expectations, however. There always were some slaves who revolted, some women who sought self-determination, some office workers who asked for access to managerial decision-making. But the power culture, seeing such things as a momentary aberration, doesn't record or provide an "institutional memory" for the efforts. Thus, the subordinate group can't find its own past accomplishments, even within a single business or institution.

Subordinates become highly attuned to the needs of the powerful people. In fact, they very often know the powerful people much better than they know themselves. (If a large part of your fate depends on accommodating to and pleasing others, you concentrate on them; there is little purpose in concentrating on yourself. This becomes even more bizarre when a subordinate person's most accessible supervisor is a machine!)

Subordinates too frequently absorb, fully believe, and internalize many of the principles, even false principles, that have been set forth by the dominant people--that is, that the subordinate is somehow inferior and really can't participate in the high level tasks and can't survive without the power holder.

Within a subordinate group, some members tend to imitate the power holders. They treat fellow subordinates as the power holders treat them, and they develop a few qualities like those of the power holders so that they gain partial acceptance by the powerful groups.

To the extent that subordinates move toward freer expression and action, they expose the inequality, raise questions about the basis for its existence, and turn hidden conflicts into open conflicts. They have to bear the burdens and take the risks that go with the "trouble-maker" label. This contradicts their entire work socialization experience and often their life-long conditioning, so subordinates, particularly women, do not do it easily.

(iii) Meanings of power. The thesis of this presentation is that in business, industry, and public institutions, managers are the power holders; clerical and secretarial personnel are the subordinate group.

You may say all that theory is interesting, but what does it mean for me? For my job? It means that a striking proportion of the work force has made only a minimal contribution to defining organizational standards for productivity, leadership, human interaction, institutional structures, values, organizational management, teamwork, and even sexual mores.

In short, power relationships must be taken into account as they affect the work life of secretarial and clerical personnel--especially when those power relationships are compounded by stereotyped expectations of power relationships between women and men.

## Looking at the future

There is some reason to think that the office of the future may become the factory of the past, particularly when one considers the potential effects of widespread automation in office work. Such a concept includes: subdivided, routinized work; high speed demands; discomforts of repetitive motions; disorientation resulting from not knowing how one's own documents, numbers, or forms fit into an integrated, whole effort; supervision by machine--the ultimate nonsupportive boss who makes a complete and continuous assessment of productivity; and demands for vigilance and concentration so that the individual cannot talk to others, or even daydream.

There are many areas that are ripe and long overdue for systematic research, but already there is sufficient evidence to say that it would be counterproductive to wait until "all the data are in" to begin to make changes. In just a few areas we have empirical evidence illustrating factors that warrant change. In other areas we have long experience, intuitive wisdom, patterns of dissatisfaction, and anecdotal evidence that say we need change. New approaches to work structure, communication processes, employee decision-making, power relationships, and program operations must be tried and evaluated--not to mention new options for physical space and environmental conditions. Employee mental health measures should be as much a part of the evaluative criteria as cost-benefit formula and traditional health and productivity measures.

Employee Assistance Programs (EAPs) are well underway throughout much of the organizational world, and health promotion and disease prevention programs are taking hold in a number of work settings. Other ancillary services such as day care centers, encouragement for participation in community activities, and family recreation opportunities are being tried in a few places. While we have a sense of psychological and emotional benefit from such programs in the work setting, we have few studies that measure their mental health impact--and none that assess the relationship of such programs to conditions in specific jobs within the organization.

Another aspect of work and mental health also merits our attention: the National Institute of Mental Health recently has launched a major educational campaign to encourage employment of mentally restored persons and to study the contribution of work to the rehabilitation of mentally ill persons. A leading edge in our collaboration with the private sector has involved the National Restaurant Association and a study of employment of mentally restored persons in food service occupations. As the program develops, our efforts will move into other enterprises where we can generate assessments of how office work influences the occupational development and emotional well-being of persons who have been mentally ill.

In summary, there is reason for concern about the psychological and emotional well-being of office workers. There is a paucity of research on the subject, and mental health measures need to be included in more studies of work life. So far most studies of office workers that address mental health do so in terms of attitudes and job satisfaction; relatively few use more detailed measures. Also, there is a need for research and demonstration projects on the role of jobs and work in rehabilitation of mentally ill persons.

REFERENCES

1 S. G. Haynes, M. Feinleib, Women, work, and coronary heart disease: Prospective findings from the Framingham Heart Study. American Journal of Public Health, 70(2):133-141, February 1980.
2 J. H. Hall, State responsibilities for the well-being of women in alcohol, drug abuse, and mental health programs. Speech presented at the ADAMHA Annual Conference of the State and Territorial Alcohol, Drug Abuse, and Mental Health Authorities, Silver Spring, Maryland, November 6, 1979.
3 T. H. Holmes, R. H. Rahe, The social readjustment rating scale, Journal of Psychosomatic Research, 11, 1967, 213-218.
4 P. Crull, The impact of sexual harassment on the job: A profile of the experiences of 92 women, Research Series Report No. 3, New York, Working Women's Institute, 1979.

# THE STRESS EFFECTS OF SEXUAL HARASSMENT IN THE OFFICE

PEGGY CRULL
Research Director
Working Women's Institute
593 Park Avenue
New York, New York

## INTRODUCTION

Over the past 8 years workers, organizers, researchers, managers, and policy-makers have been discovering and publicizing many sources of stress in what had always appeared to be one of the less dangerous and injurious occupations--clerical/secretarial work. During that same period of time, a woman named Diane Williams, who worked for the Department of Justice, was engaged in a court battle because her boss had fired her for ignoring his romantic overtures, and another woman named Carmita Wood, who worked for Cornell University, was attempting to get unemployment because her supervisor's insistent advances had forced her to leave her job. These and other cases led to the founding of a number of groups such as Working Women's Institute and the Alliance Against Sexual Coercion to study and help women combat the problem of sexual harassment on the job. Through this work, it was becoming apparent that sexual harassment was yet another hidden source of stress for women workers.

According to available surveys, sexual harassment on the job is a widespread problem affecting working women in all types of occupations and not a difficulty faced by office workers alone. One of the earliest surveys, conducted by Redbook in 1976, found that 88% of 9,000 female readers reported some form of unwanted sexual attentions in their work lives (ref. 1). A 1980 survey of government workers in Illinois showed that 59% of the women had been harassed on their jobs, and in a massive study of Federal government employees, 42% of the women told of sexual harassment within the last 2 years of their Federal service (ref. 2, 3). A telephone interview study of 139 unskilled female auto workers revealed that 36% of the interviewees had experienced sexual harassment (ref. 4). A small survey of coal miners revealed that 53% of the women had been propositioned by a boss, and 76% by co-workers; moreover, 17% had been physically attacked (ref. 5).

This paper concentrates on four areas: (1) describing the ground-breaking work of the Working Women's Institute on stress symptoms that grow out of sexual harassment on the job; (2) discussing why these symptoms are almost inevitable under these circumstances; (3) laying out some possible solutions to the problem; and (4) considering some ways in which future research on sexual harassment and stress might be done, research which will contribute to those solutions.

STRESS SYMPTOMS IN CASE HISTORIES OF SEXUAL HARASSMENT

During the 8 years it has been in existence, Working Women's Institute has had contact with thousands of women who have faced some form of sexual harassment on the job. Woven into the letters we received and counseling sessions we conducted were complaints about being afraid to go to work, feeling too tired to think, drinking too much, and fighting with co-workers. Following the lead provided by these spontaneous reports, we decided to investigate the possibility that sexual harassment acts as an occupational health hazard through the stress it creates. We wanted to know the circumstances under which the stress symptoms arise and the types of symptoms that are produced. The figures discussed here are based on case material from two sources:

1. Late in 1978 we mailed questionnaires to 325 women who had written us for assistance between 1975 and 1978. The questionnaire asked for information on the woman's age, occupation, and marital status at the time of the harassment experience. Respondents were asked to describe their experience and any effects it had on their job performance, psychological well-being, and physical health. Ninety-two completed questionnaires, or about 28% of those sent out, were returned and used in the analysis.

2. The second source of material is records kept on clients of the Institute's crisis counseling service. Except for occupation and description of the sexual harassment situation, the information contained in the mailed questionnaire is not necessarly asked for or recorded in counseling sessions. However, much of that information arises spontaneously in the sessions and is noted in the counselor's records. The present analysis used records of the 170 clients who had complete counseling sessions at the Institute between February and December of 1979.*

The women in this sample come from every kind of work setting and every sector of the economy. The distribution of occupations in our sample is similar to that of all women in the work force except that clerical workers are more heavily represented. Forty-six percent of the women we hear from are

---

*The Institute has similar information on 250 additional clients. An analysis of those cases appears in Crull and Cohen (ref. 6). Crull (ref. 7) is a summary of data from the questionnaire survey and the 1979 and 1980 study. The results on the questionnaire study are reported in Crull (ref. 8), and a brief report of the present results appears in Crull (ref. 9). The percentages in this report differ slightly from the 1982 report because some of the coding categories were refined and the data reanalyzed.

clerical/secretarial workers, whereas about 34% of the female work force is clerical. There are several possible explanations for this over-representation of clerical workers. It is quite likely that this figure is based on the high proportion of clerical workers in the New York metropolitan area. However, it could be that clerical workers are more frequently targets of sexual harassment or that they are more willing to report it when it happens.

Before discussing the types of stress we have uncovered, sexual harassment should be defined. At the Institute we have developed the following definition:

> Sexual harassment is any attention of a sexual nature in the context of work which makes a woman uncomfortable, impedes her ability to do her work, or interferes with her employment opportunities. It can be in the form of comments, looks, touches, sexual epithets, propositions, even rape at work.

We are not just talking about losing a job for not sleeping with the boss. We are talking about a whole range of attitudes from being expected to look sexy, to being turned into the butt of sexual jokes by customers or co-workers, to being coerced into having an affair with a superior. This whole range of behaviors and the circumstances that surround them are the sources of the stress discussed in this paper.

In order to look at that stress, we divided the effects reported by our subjects into three categories: (1) psychological effects; (2) physical effects; and (3) effects on work performance and attitudes. Under each of these categories, a number of specific effects were mentioned. Using the categories of specific effects, we classified and summarized the information in the questionnaires.

We first looked at the effects on the woman's psychological well-being. Ninety-two percent mentioned such effects. The major complaint was that they experienced excessive tension (61%) that lasted well beyond working hours. The following quote reveals what that tension is like:

> I was nervous all day at work to the point where I fouled up when I was changing the paper in the Xerox machine because that obnoxious salesman came to stare at me while I bent over. I got to the point where I couldn't relax at home after a day of putting up with him and his buddies. I asked my husband to give me some of his Valium, but he said I should quit instead.*

Many women reported that the constant tension caused them to increase their consumption of alcohol and tranquilizers.

---

* In order to assure confidentiality, all quotes used in this paper are adaptations or composites of statements from people in the sample.

The next most common symptoms were anger (34%) and fear (19%). Women said they were afraid to go to work, afraid to tell their families, afraid to complain. Anger, which could be a useful tool in helping a woman combat the sexual harassment, was often so excessive that it became dysfunctional. It overflowed in many cases to the very people who might be able to help resolve the situation such as personnel, union officials, or intake workers at unemployment. It was sometimes taken out on husbands, children, or friends rather than on the harasser or uncooperative company officials.

About 36% of the women in our sample pointed out physical ailments they thought had been brought on by their sexual harassment dilemma. Leading this list were nausea (13%), headaches (11%), tiredness (5%), and drastic weight changes (5%). It was not unusual for the women to fail to make the connection between the sexual harassment situation and the physical symptoms at first. For example, a computer programmer went to the doctor for severe migraine headaches. It was not until the doctor began questioning her about her job that she realized that the headaches were correlated with regular propositions from her boss. About 12% of the women eventually seek help from a doctor or therapist to alleviate the symptoms.

Finally, we looked at the direct effects sexual harassment had on work performance and attitudes. Even though many were proud of having maintained their ability to work productively, 46% of the women reported that their performance and attitudes were negatively affected by their experience. The major complaint was that it was difficult to concentrate (17%) on tasks at hand. One receptionist said that she had trouble answering the phone whenever a certain salesman was around because she was made uncomfortable by his stares and suggestive remarks. Many women (9%) lost work time trying to figure out ways to avoid the harasser. Carmita Wood, whose unemployment case was mentioned earlier, developed such avoidance techniques as dressing in an austere manner and taking circuitous routes to and fom her desk. Other women began taking extra long lunch hours or more frequent sick days.

The direct effects of the harassment on attitudes toward work go far beyond the immediate situation. The inability to focus fully on the job and the ambiguity of the relationship with a boss or co-worker that is infused with sexual innuendoes begin to eat away at a woman's self-confidence (11% reported this effect). The result was that their motivation to go to work or even to pursue their careers waned because they felt they were not suited to the environment in which they found themselves.

## WHY SEXUAL HARASSMENT RESULTS IN STRESS SYMPTOMS

The discovery of these stress symptoms in women who have experienced sexual harassment is not surprising given the nature of the incidents, and the economic and social circumstances in which working women function. The profile of the sexual harassment situation itself helps to explain the appearance of stress. The incidents described by the women in our sample are not the simple chase around the desk depicted in popular literature. Instead, they consist of persistent, pervasive, or physically intrusive behaviors. The advances may persist for a long time without letting up. Doreen Romano, a bookkeeper who won a settlement in New York State, was fired by her boss who had been propositioning her for 2 years. Often the harassment is pervasive in the sense that more than one man is involved or the whole office ritual and atmosphere is filled with sexual joking and innuendo in which women employees are expected to participate, even though much of the content derogates them. In about 18% of our cases the harassment is much more physically intrusive and even intimidating than one might imagine. About half of the women in the group we studied were touched or grabbed in a sexual manner and 20% of them were subjected to physical force.

There are some special features of the profile of sexual harassment for clerical/secretarial workers that add to its potential for producing stress. From our data it appears that women in clerical/secretarial jobs are ordinarily subjected to the type of harassment that at first seems like harmless flirting or propositions but begins to cause difficulty because it is unremitting or connected with some sort of threat. In a recent article in the Journal of Legislation, Judy Ellis has labeled this "exploitative harassment" in contrast to what she calls "generalized harassment" where the woman is intimidated, ridiculed, or derogated by the use of lewd remarks, sexual epithets, and pornographic pictures (ref. 10). Secretaries encounter this second type less frequently than women in nontraditional jobs who are often seen as intruders by their co-workers (ref. 11). Therefore, they are likely to feel they should be flattered by the attention they are receiving. When it becomes too much of an interference or turns to threats, the woman who has been the focus of the attention may begin to feel guilty or confused or feel that she has no right to stop the behavior because, "after all, it is only meant as a compliment." In short, the ambiguity of the situation may be even more stressful than a clearly negative situation. The large percentage of clerical workers who are minority women have a double burden when they encounter sexual harassment because it is frequently mixed with racial harassment. This is especially true for black women who have historically been depicted as sexually dangerous (ref. 10). Women of color who have come to the Institute for assistance frequently report that it was some time before they were able to distinguish that they were the targets of <u>two</u> kinds of harassment rather than just the one, racial harassment, with which they were familiar.

Another reason that harassment causes stress is that it does not typically stop with sexual overtures but is followed by something we have come to call "work harassment." As soon as it becomes clear that the woman is not going along with the overtures or plans to complain about them, she is met with a barrage of undue criticism of her work; she is asked to work overtime when it isn't necessary; she is ridiculed in front of co-workers or clients; or she gets negative evaluations when her work was perfectly acceptable a few months before. For more than a quarter of our sample the end result was that they were fired or laid off. Another one-third of the group were forced to resign from their jobs because of the pressure of the sexual and work harassment.

To add insult to injury, the harassment often continues even after the woman's job ends. She is met with disbelief from officials at unemployment or the EEOC or told that she doesn't have a strong enough case or treated unsympathetically.

In order to understand fully the appearance of stress symptoms in these cases, we must take a look at the economic and social context in which they occur. Simply put, the major economic reality surrounding sexual harassment is that women need their jobs. Very few women work for "pin money" anymore. They are either heads of their households or contributors of an income that is essential for family survival. Unfortunately, most women hold lower status jobs than the men with whom they work and, in fact, are often supervised by men (ref. 12). In addition, they are less likely to be in unions than men, and their work is usually defined as less skilled and valuable than that of men (ref. 13). All of these characterizations are especially true of clerical/secretarial work, and they all add up to the fact that the man who sexually harasses a woman is likely to have some form of power or control over her job. Even a man, who by all of these measures of status and power is in an equal position with a woman, may have more "pull" than she does by virtue of informal networks around the office in which men are more likely to participate. In our sample, 80% of the harassers had direct or indirect authority over the woman's job. In other words, a woman's harasser is likely to win out in any situation where he decides to threaten her job.

The socialization that males and females receive while growing up adds to the difficulties created by the economic situation. The culture teaches men to see sexual assertion and success as signs of masculinity (ref. 14, 15). It also teaches them that women are supposed to show their sexual interest in a discreet manner. On the other hand, women learn that they are responsible for both male sexual behavior and their own. This means that the woman's attempt to stay away from the harassment can go wrong in many ways. Her refusal can easily be seen as a put-down or a challenge. If she tries to save the ego of the man or men who made the advances by politely ignoring them, her actions may be mistaken for a discreet "yes" and the advances may then become more insistent. Then she is back where she started with the dilemma of having to refuse more directly. In any case, her job is in jeopardy because of the escalating misunderstandings that are built into sex-role socialization.

Two other aspects of socialization come into play here and create stress. Being thrown into this situation often exacerbates conflicts within the woman and among her family members about working outside the home. Sexual harassment may suggest to her that "her place is in the home." Many women attempt to escape the guilt and conflict by looking for a job where they will come into contact with few men, or they begin to lose interest in working altogether. Of course, most women don't actually leave the work force as a result of a sexual harassment experience because they don't have that luxury. The other problem that sexual harassment often restimulates for women is lack of self-confidence about their skills that many of them learn from childhood. The work harassment that often follows sexual harassment may make the woman unable to judge whether she is doing her job well, and this is especially true if she has come into the situation with shaky self-confidence.

These results and interpretations need to be qualified in two ways. First, we do not know how typical these stress reactions are since we don't know if this sample is typical of women who experience sexual harassment. The fact that the women in the sample have sought information, referral, and counseling gives us some clues about how their situations compare with those of other women. Their experiences are probably more serious than those of the general population. For example, women whose situations are resolved quickly and without extensive repercussions are not as likely to write for help or seek counseling. We can speculate that they do not experience the high level of stress found in our sample. On the other hand, many women too isolated to know where to turn or too ashamed to talk to anyone about their experience probably do not contact us. We have heard of cases in which women, desperate to hold onto their jobs and unaware of any alternatives, have submitted unwillingly to sexual advances, sometimes including intercourse, for a period of months or years without telling even their closest friends. This situation undoubtedly leads to more severe reactions than those we have described in our sample. It should also be noted that our sample does not include women who have been engaged in mutual sexual behavior such as reciprocal flirting, dating, or an affair that has been agreed upon by both parties and which had no negative consequences for their work. There is no reason such women would seek help. Our sample _does_ include several cases of women who became targets of work harassment or received disparate and discriminatory treatment after a mutual relationship with a boss or supervisor ended.

The second qualification of these results is that they are exploratory. Our content analysis grew out of the salient themes in our cases and we did not attempt any tests of significance. Although our results have been supported by research with other samples (ref. 1, 3, 4), no study has used statistical techniques that establish a causal relationship between sexual harassment and stress. Gruber and Bjorn (ref. 4), however, attempted to correlate frequency of harassment with levels of stress and found that frequent harassment was associated with low self-esteem and diminished general life satisfaction. At this point, our results and those in similar studies need to be extended by further research.

## SOLUTIONS TO SEXUAL HARASSMENT ON THE JOB

There is a wide variety of possible solutions to this problem, ranging from the personal to the societal. The goal is not simply to treat the stress but to alter the conditions that create it.

Some of the groups that have formed to act as advocates for women who are undergoing sexual harassment have turned their attention to developing short-term coping solutions. This is a response to the overwhelming number of requests that these groups receive for immediate assistance. Groups such as Working Women's Institute, the Alliance Against Sexual Coercion in Cambridge, and the Coalition Against Sexual Harassment in Minneapolis have developed counseling and workshops aimed at helping women head off sexual harassment and deal with the repercussions, both economic and psychological, of situations that they have been unable to prevent.

More permanent prevention and protection are being sought through changes in the law. The major law that has been used to help women gain their rights in sexual harassment situations is Title VII of the 1964 Civil Rights Act. The courts have ruled that sexual harassment is considered discriminatory. These guidelines put employers on notice that they are responsible for creating a workplace free of sexual harassment. Other laws, many on the state level, have been used to remedy the damage caused by sexual harassment--tort law, unemployment, and workers' compensation.

Of course, we would hope that women will not have to go to court to obtain respect and just treatment at work. To that end, other mechanisms have been developed to rectify the sexual harassment before it forces the woman to go to outside sources for help. Many unions have begun to include sexual harassment in their contracts and have considered special grievance procedures for such cases. Among the pioneers in this area are the United Auto Workers and the American Federation of State, County, and Municipal Employees. Many employers have formulated policy statements condemning sexual harassment by their management and supervisory personnel. Others have sponsored workshops that give their employees an opportunity to learn about both personal and institutional issues surrounding sexual harassment.

None of these efforts can result in any permanent change without vigilant grassroots involvement. At the Institute we have been in touch with many groups of working women and men who have organized at their workplaces to stop sexual harassment and support women who have encountered it. For example, as a result of several incidences of sexual harassment, a small group of women has formed a committee to talk about women's issues at a large northeastern newspaper. They are hoping to discuss such problems as discrimination in pay and promotions as well as sexual harassment. On the west coast the California Rural Legal Assistance Workers was founded specifically to deal with a situation in which a lawyer grabbed the breast of the receptionist with whom he worked.

## FUTURE RESEARCH ON SEXUAL HARASSMENT

More research needs to be directed to this problem in order to support these efforts at eradicating sexual harassment as a source of stress in the workplace. First, as mentioned above, we need more controlled research in which, for example, groups of harassed and nonharassed women can be compared for levels of stress. Further, we need to know if and how different occupations generate different types of sexual harassment and if those different types cause variations in stress levels. Once these causal connections are established and variations explored, data on the possible long-term effects must be gathered. Information such as this will not only have theoretical value, but will be useful in demonstrating to policy-makers the need for preventing sexual harassment. For example, our data on the stress effects have been central in making the legal argument that sexual harassment is discriminatory even when it does not result in the loss of a job, raise, or promotion. We have testified to the fact that a woman who is under stress is being subjected to an unequal condition of employment based on her sex. We have also testified to the seriousness of these stress effects in order to help determine what damages women should receive. Other researchers have been looking at stress with an eye to policy. Recently the Merit Systems Protection Board calculated the loss to taxpayers that is caused by Federal employees taking sick days as a result of sexual harassment (ref. 3).

Although little attention has been given here to individual solutions, it should be emphasized that much more research is needed to give immediate relief to sexual harassment victims. One way to provide this relief is to investigate which education and counseling techniques are most effective. For instance, in our own research at the Institute we have discovered that one major problem in counseling is that women often wait until they have lost their jobs before seeking any outside assistance. At present we are attempting to find out why women do not recognize that they have a problem before they lose their jobs.

Research can be used to determine which solutions are working. Although all efforts at resolving the problem are valuable at this moment, it would be useful in the future to compare the effectiveness of various policies, laws, and procedures in preventing sexual harassment. But the final message of this paper must be that we will not be able to eliminate sexual harassment altogether until the unequal status between men and women in the work force is rectified.

REFERENCES


1 C. Safran, What men do to women at work, Redbook, 1976.
2 B. Hayler, Testimony before the House Judiciary II Committee, State of Illinois, 1981.
3 Merit Systems Protection Board, Sexual harassment in the Federal workplace: Is it a problem? A report of the U. S. Merit Systems Protection Board, Office of Merit Systems Review and Studies, 1981.
4 J. Gruber and L. Bjorn, Blue collar blues: The sexual harassment of women autoworkers. Work and Occupations, 9:3 (1982) 271-298.
5 C. White, B. Angle and M. Moore, Sexual harassment in the coal industry: A survey of women miners, unpublished mimeo. Coal Employment Project, Oak Ridge, Tennessee, 1981.
6 P. Crull and M. Cohen, Expanding the definition of sexual harassment on the job. Working Women's Institute, Research Series, Report No. 4, 1982.
7 P. Crull, Sexual harassment, stress, and women's health, in W. Chavkin, Report from the Front Lines: Occupational and Environmental Health Hazards Confronting Women, Monthly Review Press, 1983.
8 P. Crull, The impact of sexual harassment on the job: A profile of the experiences of 92 women, in D. Neugarten and J. Shafritz (Eds.), Sexuality in Organizations: Romantic and Coercive Behaviors at Work, Moore Publishing Company, Oak Park, Illinois, 1979, pp. 67-71.
9 P. Crull, The stress effects of sexual harassment on the job: implications for counseling, American Journal of Orthopsychiatry, 52:3 (1982) 539-544.
11 J. Ellis, Sexual harassment and race: a legal analysis of discrimination, Journal of Legislation, 8:1 (1981) 31-45.
11 P. Crull, Sexual harassment and male control of women's work, Women: A Journal of Liberation, 8:2 (1982) 3-7.
12 U. S. Department of Labor Employment Standards Administration, Women's Bureau, The earnings gap between women and men, 1979.
13 Women and Work Study Group, Understanding the problems of women at work, unpublished mimeo, New York, 1981.
14 L. Phelps, Female sexual alienation, in Jo Freeman (Ed.), Women: a Feminist Perspective, Mayfield Publishing Company, Palo Alto, 1979, pp. 18-26.
15 M. Rosenbaum, Clarity of the seduction situation, in J. Wiseman (Ed.), The Sociology of Sex, Harper and Row, New York, 1976, pp. 51-57.

# IMPACT OF WOMEN'S WORK ON FAMILY HEALTH

Chaya S. Piotrkowski,
Department of Psychology
Institute for Social and Policy Studies
Yale University
New Haven, Connecticut 06520

## INTRODUCTION

Since World War II, a subtle revolution has occurred in women's employment. Not only are more women in the economy, but the pattern of their employment also has changed: increasing numbers of married women with school-aged and preschool children have entered the labor force. In 1979, 59% of married mothers of school-aged children were employed, as were 43% of married women with preschoolers (ref. 1). Of course, women have always worked and contributed to the subsistence of families. It is a relatively modern phenomenon that most women now have two physically separate and socially distinct primary work roles--one in the family as a household worker and one outside it as a wage earner.

A number of studies have discussed the effects of paid work on women's health. Depending on which health indicators one examines, different patterns of outcome emerge. This paper focuses on the effect of women's employment on *family health*. Family health is conceived of broadly to include the physical and emotional well-being of individual family members, their adequacy in performing family roles, and the degree of satisfaction family members derive from their relationships. A focus on family health is useful because it allows us to assess both the direct and indirect consequences of particular working conditions. Two aspects of family health that are important to women workers are addressed in this paper: the quality of their relations with children and husbands, and the consequences to women of balancing family and employment responsibilities.

Although we speak of effects and impacts, it is important to note that we are only dealing with statistical associations between variables. Nevertheless, certain causal relations are more plausible than others and, historically, work life appears to have influenced families more than vice versa. Unfortunately, research on employment and family life has not proceeded far enough for us to specify the effects of various occupational conditions on family health. We know much less about the impact of occupational conditions on families than on individual well-being. Research on the work and family life of women wage earners in nonprofessional and nonmanagerial occupations, however, has relevance to secretarial and clerical workers. We can assume that clerical workers constitute a substantial proportion of such research samples.

Understanding the relationship between work and family life is important for both men and women. Since men are not viewed primarily in their family roles, their work-family relations have received less public attention. Terms such as "working mothers" or "working women" can be misleading because they obscure women's household work role. Unless they are wealthy, all mothers are working mothers, especially if they have preschoolers at home. We should remember that many women still work primarily in the household and that employed women also still do the lion's share of housework.

Despite a substantial literature on women's employment and family life, little is known about the relationship between specific occupational conditions and family well-being. Social and behavioral scientists have proceeded as if only the fact of women's employment were important; that is, whether or not women are in the labor force. Few researchers studying women's work and their family life have considered the particulars of women's jobs. In other words, few studies examine factors such as quality of supervision, salary, job security, job satisfaction, number of hours worked, etc., variables that commonly are studied for men. The emphasis on the fact of women's employment rests on the assumption that it is primarily women's absence from the home that is important, not what they do on the job nor the conditions under which they do it. This neglect of concrete conditions of employment stems from the social and political context in which this research occurs, namely, the debate over women's "proper place." In this context, researchers have asked two major questions: Does women's employment harm children? Does women's employment harm marriages? Consequently, there are numerous studies comparing the children of employed and nonemployed women and the marriages of dual- and single-earner couples. Unfortunately, these studies tell us very little about the ways in which the specific conditions of work may affect families. It is important to move away from the debate about women's proper place. Women are and will continue to be employed in large numbers. Their wages are necessary to support themselves and their families. We shouldn't be asking simply if the fact of their employment harms families, but how different conditions of work may influence family life.

## REVIEW OF RESEARCH ON WOMEN'S EMPLOYMENT AND FAMILY HEALTH

There is no strong evidence that women's employment per se harms children. In fact, it may benefit them--particularly girls (ref. 1).

The children of employed women may have different values and orientations than the children of nonemployed women. For example, the children of employed mothers see women as more competent and have more egalitarian sex role attitudes than the children of nonemployed women (ref. 2, 3). Of course, here causal relationships are especially difficult to untangle.

A fascinating question that we are beginning to research is how specific job conditions shape the behavior of children. Melvin Kohn and his colleagues at the National Institute of Mental Health (ref. 4) demonstrated that certain occupational conditions, such as the degree of self-direction permitted by a job, influences people's personalities and their child-rearing values. In our research we have hypothesized, for example, that parents who lack the opportunity to utilize their training on the job will not encourage skill acquisition in their children (ref. 5). Our findings support this hypothesis. Using objective measures, we have found significant associations between the extent of women's skill utilization at work and their children's classroom math achievement. Skill underutilization is associated with lower math achievement, even when mothers' education is statistically controlled. We believe that occupational conditions serve to socialize workers in ways that affect other family members. Such findings suggest the subtle effects that occupational conditions can have on women and their families. They have special relevance for the families of clerical workers who often find themselves in routine, boring jobs.

If we turn briefly to mother-child relations, we find that infants are just as attached to and older children feel just as close to employed as nonemployed mothers (ref. 6, 7, 8). In our own research, we found that adolescent girls in a minority sample felt closer to their mothers as the hours mothers spent in employment increased (ref. 9). Adolescent girls may appreciate the independence given them by mothers who are employed or may simply admire their mothers' contribution to the family wage through employment. In either case, children don't seem to experience their mothers' employment as rejection nor is extent of labor force participation--within extremes, of course--clearly detrimental to the mother-child relationship.

The idea that a mother's noncontinuous presence in the family would harm children is based on a mythology of full-time mothering as some wonderful condition in which women are constantly available to their children. It is this mythology, buttressed by methodologically questionable research, that resulted in the equation of mother's employment with maternal deprivation. Actually, we know almost nothing about what does occur between full-time housewives and their children (ref. 2). Personal research suggests that a considerable amount of their time is spent in struggle (ref. 10). As Lois Hoffman has noted, it is not at all clear that nonemployed mothers actually spend more time in attentive, positive interaction with their children than nonemployed mothers. In fact, several studies suggest that employed mothers take special pains to spend quality time with their children (ref. 11, 12). We have found that regardless of the number of hours mothers were employed, their adolescent daughters reported no differences in how available and supportive their mothers were to them (ref. 13).

There is no consistent evidence that men married to employed women suffer poorer health status than the spouses of nonemployed women (ref. 14). And whether one asks husbands or wives, in most cases the marriages of dual-earner couples seem to be no less happy than those of traditional couples in which the husband is sole breadwinner. Although some studies have found greater conflict in dual-earner marriages, disagreements tend to be local and focus on domestic matters (ref. 15). It is important that we do not unquestioningly assume that such disagreement is necessarily unhealthy. It also may reflect more equalitarian decision-making processes (ref. 16).

While it is true that among white women, at least, employment rates are higher among divorced than married women, cross-sectional studies cannot tell us what came first: the divorce or the job. Employed women more often report contemplating divorce than nonemployed women, even though they are equally satisfied with their marriages (ref. 17, 18). This finding is a fairly reliable one. Does that mean that employment "causes" divorce? Not at all. One longitudinal study did suggest that a woman's actual or potential earnings relative to her husband's slightly increased the probability of divorce (ref. 19). But the effect was very small and it was not employment _per se_ but actual or potential earnings that made a difference. These results suggest the operation of what has been called an "independence" effect. Insofar as women's employment increases their economic independence, they may be more willing to consider alternatives to marriage and to act on those alternatives when their marriages are unsatisfactory. Marriage may simply be less important to their overall satisfaction (ref. 20). This is a far cry from saying that employment causes divorce.

There does exist an interesting paradox that has not been explored. If we consider only women with a high school education or less, we find an apparent contradiction: Employed women report less depression and higher self-esteem (ref. 21, 22), but lower marital satisfaction than nonemployed women (ref. 23, 24). Simple group comparisons do not tell us much about the _processes_ underlying this paradox. Perhaps the independence and self-esteem gained through employment allows women to evaluate their marriages less defensively or leads them to expect more from their husbands. In such a case lowered marital satisfaction does not necessarily imply a negative outcome. Our inability to account for such consistent effects (this finding about lowered marital satisfaction is fairly reliable) indicates the poverty of approaches that simply compare groups of employed and nonemployed women.

Employment _per se_ does not appear to be clearly related to family health. Those few studies that have gone beyond the fact of employment indicate that degree of satisfaction or dissatisfaction with one's work may be a more important determinant of family health than the fact of employment, at least for mother-child relations. Two types of satisfaction have been considered: satisfaction with one's work role, and satisfaction with one's particular job. Women who are in their preferred work role tend to have more satisfying relations with their children than women who are not. That is, it seems to be important for family relations that women are doing what they prefer. One study found that the most inadequate mother-child relations existed for full-time housewives who wanted to be employed but were not (ref. 25). These women felt duty-bound to stay at home. Other studies suggest that these frustrated women are overprotective of their children, are threatened by their children's independence, feel anxious, and have low self-esteem (ref. 2).

Daily working conditions that affect womens' job satisfaction and morale also may influence mother-child relations. Several studies support this hypothesis (ref. 26, 27). We have found that the more positively women feel on the job, the more supportive and available they are to their daughters (ref. 13) (in press). Our data indicate that morale is improved by jobs that are interesting and provide some autonomy. These results have special implications for clerical workers who often are in demanding, relatively uninteresting jobs with little autonomy. Such jobs may have costs both for women and their children. Some theorists have argued that interesting, satisfying jobs may cause problems for families (ref. 28). They reason that an interesting job causes a person to invest most of himself or herself in the job to the neglect of the family. Taking that argument to its extreme, of course, suggests that uninteresting jobs are best for the family. However, there is little evidence to support this claim. If one has a satisfying job, family relations probably will benefit.

The small, but consistent, body of research on job satisfaction and mother-child relations suggest that the social costs of unsatisfying and stressful jobs are broader than we usually think. That is, they may affect those people--such as children--who are not directly involved. We currently are conducting research specifically addressing this question. We are attempting to determine whether job stress affects the families of workers, including their children. The model we are using is one developed by Robert Karasek. He has found that the combination of being in a job with high demands while simultaneously having little job control was related to signs of physical and psychological strain (ref. 29). We are extending that model to consider strain in the families of such workers. We hope to learn more about the indirect health costs to families of stressful and unsatisfying jobs.

Women consistently complain about difficulties in juggling family and employment responsibilities. How do they do it? Are they all superwomen? We have few answers to these important questions. We do know that employed women spend less time doing housework than nonemployed women (ref. 30). There is some debate about whether the husbands of employed women do more housework than the husbands of nonemployed women. Even the studies that indicate that they do find that the increment is small (ref. 31). In other words, a little difference may make little difference. Findings from a recently reported study indicate that husbands participate more in housework when their wives' earnings are high relative to theirs (ref. 32). Professional and managerial husbands, by this account, do the least amount of housework.

The more roles people have, the more conflict they report because of simultaneous role demands (ref. 33). Feelings of role strain resulting from conflicts and overload are more intense for employed women than men and increase with hours worked (ref. 34). For women, problems in scheduling job and family tasks contribute to such difficulties (ref. 35). Flexible working schedules may be an important innovation for reducing work-family management strain for women. However, it should not be considered a panacea. Extent of such flexibility is important; one recent study found that women with children did not benefit greatly from modest work scheduling (ref. 37) or flexibility (ref. 36). More important may be job autonomy more generally considered. Flexibility in work schedules is but one alternative that should be considered to help women minimize role overload. Although part-time employment for women also has been advocated as a partial solution to role overload, such a strategy may have hidden financial costs for families (ref. 10).

How, then, do employed women manage? One individual solution to minimize work-family management strain is to create careful priorities (ref. 12). Women may reduce standards for housework and give priority to children, allocating scarce time carefully. They may sleep less or at least reduce their personal leisure time. In a recent General Mills Report (ref. 38) on work and family, two-thirds of employed women sampled expressed dissatisfaction with the amount of personal time they had. Only one-third of men were similarly dissatisfied. Again, at least one study indicates that, for individual health, degree of satisfaction with one's roles is more important than the number of roles held (ref. 39). Job satisfaction, therefore, may minimize potential strain associated with role overload. One of the special difficulties women face is that there is no special zone that is work-free for them. While their husbands may spend long hours at work, for them home can serve as a "haven" free of work. For women, such a haven may not exist, since home is also a workplace (ref. 10). For example, employed women spend more time in housework on weekends than men do. They simply have no place to get away.

As is evident from this synopsis of a limited segment of the research literature, we are just beginning to learn how employment may influence family well-being. Unfortunately, much existing research on women's employment and family relations has been based on samples of married, white women. Career women and dual-career families also have received a disproportionate amount of attention. Most women do not have careers and never will; most women have jobs. We need more careful research into the nature of these jobs. We also know almost nothing of the dilemmas of managing work and familylife in female-headed households and in minority families (ref. 37). Rather than continuing to ask whether the fact of women's employment is detrimental to children or marriage, research efforts need to be directed at describing the conditions under which specific features of women's employment roles and the quality of work life influence their families. If it is true, as we suspect, that stressful and unsatisfying jobs may have special costs for women and their families, then clerical workers and their families represent a relatively high risk group. By including the families of workers in our health assessments, we are in a better position to evaluate the full consequences of the working conditions of both women and men. A clear understanding of such consequences will help us identify groups at risk and policies that maximize well-being.

REFERENCES


1 L. W. Hoffman, The effects of maternal employment on the academic attitudes and performance of school-aged children. School Psychology Review, 9 (1980) 319-335.
2 L. W. Hoffman, Effects on child in L. W. Hoffman and F. I. Nye (Eds.), Working Mothers, Jossey-Bass, San Francisco, 1974, 126-166.
3 L. W. Hoffman, Maternal employment. American Psychologist, 34 (1979) 859-865.
4 M. L. Kohn and C. Schooler, The reciprocal effects of the substantive complexity of work and intellectual flexibility: longitudinal assessment. American Journal of Sociology, 84, (1978) pp. 24-52.
5 C. S. Piotrkowski and M. H. Katz, Indirect socialization of children: the effects of mothers' jobs on academic behaviors. Child Development (1982) 53 1520-1529.
6 C. Etaugh, Effects of material employment on children. Merrill-Palmer Quarterly, 19-20 (1974) 71-98.
7 A. M. Propper, The relationship of maternal employment to adolescent roles, activities, and parental relationships. Journal of Marriage and the Family, 34 (1972) 417-421.
8 E. T. Peterson, The impact of maternal employment on the mother-daughter relationship. Marriage and Family Living, 23 (1961) 355-361.
9 C. S. Piotrkowski and M. H. Katz, Work hours, the quality of work experience and mother-daughter relations, Department of Psychology, Yale University, unpublished manuscript, 1981.
10 C. S. Piotrkowski, Work and the Family System, The Free Press, New York, 1979.
11 R. Rapoport and R. Rapoport, Dual-career families re-examined, Harper Colophon Books, New York, 1977.
12 A. D. Harrison and J. H. Minor, Interole conflict, coping strategies, and satisfaction among Black working wives. Journal of Marriage and the Family, 40 (1978) 799-805.

13 C. S. Piotrkowski and M. H. Katz, Work experience and family relations among working class and lower middle class families, in H. Z. Lopata and J. Pleck (Eds.), Research in the interweave of social roles, Vol. 3; Families and Jobs, Jai Press, Greenwich, in press.
14 A. Booth, Wife's employment and husband's stress: a replication and refutation. Journal of Marriage and the Family, 39, (1977) 645-650.
15 A. Gianopulous and H. E. Mitchell, Marital disagreement in working wife marriages as a function of husbands' attitude toward wife's employment. Marriage and Family Living, 19 (1957) 373-378.
16 J. S. Shasetz, Resolution in marriage: toward a theory of spousal strategies and marital dissolution rates. Journal of Family Issues 1 (1980) 397-421.
17 F. I. Nye, Marital interaction, in F. I. Nye and L. W. Hoffman (Eds.), The Employed Mother in America, Rand McNally, Chicago, 1963.
18 A. Booth and L. White, Thinking about divorce. Journal of Marriage and the Family, 42, (1980) 605-616.
19 A. Cherlin, Work life and marital dissolution, in G. Levinger and O. C. Moles (Eds.), Divorce and Separation, Basic Books, New York, 1979.
20 E. Haavio-Mannila, Satisfaction with family, work, leisure, and life among men and women. Human Relations, 24 (1971) 585-601.
21 M. M. Ferree, Working class jobs: paid work and housework as sources of satisfaction. Social Problems, 23 (1976) 431-441.
22 L. Radloff, Sex differences in depression: the effect of occupation and marital status. Sex Roles 1 (1975) 249-265.
23 G. L. Staines, J. H. Pleck, L. J. Shephard and P. O. O'Connor, Wives' employment status and marital adjustment. Psychology of Women Quarterly 3 (1978) 90-120.
24 F. I. Nye, Husband-wife relationship, in L. W. Hoffman and F. I. Nye (Eds.), Working Mothers, Jossey-Bass, San Francisco, 1974, 186-206.
25 M. R. Yarrow, P. Scott, L. DeLeeuw and C. Heinig, Child-rearing in families of working and nonworking mothers. Sociometry 25 (1962) 122-140.
26 J. Harrell and C. Ridley, Substitute child care and the quality of mother-child interaction. Journal of Marriage and the Family, 37 (1975) 556-564.
27 L. W. Hoffman, Mother's enjoyment of work and effects on the child, in F. I. Nye and L. W. Hoffman (Eds.), The Employed Mother in America, Rand McNally, Chicago, 1963, 95-105.
28 M. Young and P. Willmott, The symmetrical family, Pantheon, New York, 1973.
29 R. J. Karasek, Job demands, job decision latitude and mental strain: implications for job redesign, Administrative Science Quarterly, 24 (1979) 285-301.
30 J. P. Robinson, Housework technology and household work, in S. F. Berk (Ed.), Women and Household Labor, Sage Publications, Beverly Hills, 1980, 53-68.
31 J. H. Pleck, Men's family work: three perspectives and some new data. The Family Coordinator, 28 (1979) 481-488.
32 S. Model, Housework by husbands. Journal of Family Issues, 2 (1981) 225-237.
33 J. B. Herman and K. K. Gyllstrom, Working men and women: inter and intra-role conflict. Psychology of Women Quarterly, 1 (1977) 319-333.
34 P. M. Keith and R. B. Schafer, Role strain and depression in two job families. Family Relations, 29 (1980) 483-488.
35 J. H. Pleck, G. L. Staines, and L. Lang, Conflicts between work and family life. Monthly Labor Review, 103, (1980) 29-32.
36 H. H. Bohen and A. Viveros-Long, Balancing jobs and family life, Temple University Press, Philadelphia, 1981.
37 M. H. Katz and C. S. Piotrkowski, Correlates of family role strain among employed Black women. Family Relations, 32 (1983).
38 General Mills Inc., Families at work: strengths and strains, Minneapolis, Minnesota, 1981.
39 E. Spreitzer, E. E. Snyder and D. L. Larson, Multiple roles and psychological well-being. Sociological Focus, 12 (1979) 141-148.

# RESULTS FROM WORKING WOMEN'S OFFICE WORKER HEALTH AND SAFETY SURVEY

JUDITH GREGORY
Working Women Education Fund
1224 Huron Road - 3rd Floor
Cleveland, Ohio 44115

## THE OFFICE WORKER HEALTH AND SAFETY SURVEY

During 1980-1982, the Working Women Education Fund in conjunction with Working Women has carried out a pilot project on clerical workers' occupational health in two cities, Cleveland and Boston. Our Project: Health and Safety is partially funded by a New Directions grant from the Occupational Safety and Health Administration. As part of the project, the Office Worker Health and Safety Survey was distributed in both cities in the fall of 1980 by Working Women's affiliate groups, Cleveland Women Working, and 9 to 5 (in Boston).

The survey was designed as an outreach and education tool. The survey asks 70 questions in the areas of job stress, office environment, health problems, and job and personal characteristics (ref. 1). It was not intended as an instrument for structured research. We used it to ascertain the main concerns of women office workers, and to get a first measure of the prevalence of problems and the range of health effects.

When we received 1,300 responses, our advisors urged us to look at cross-tabulations by computerizing the information. Although these observations are clearly impressions rather than hard data, I hope they will prove useful. The results from the 960 responses are described herein.

### How the survey was conducted

Only 11% of all women clericals in the United States are represented by unions today. In private industry, estimates run as low as 6%. There are two points to keep in mind relevant to this lack of unionization: (1) the vast majority of clerical workers have very little influence over their working conditions, and (2) because there are difficulties in reaching unorganized office employees, they are under researched as a group.

To carry out the survey, we distributed 8,000 copies in each city. We handed out surveys early in the morning at major public transit stops and in front of downtown office buildings, and at noon at public gathering places and the busiest street corners. The surveys were returned by mail post paid. Each respondent had the option of giving her name or maintaining her anonymity. We have found this to be a very effective method of reaching unorganized clerical workers. In addition to tabulating the results from the surveys, follow-up interviews were conducted with approximately 100 respondents with a variety of problems and concerns.

Job and personal characteristics of survey respondents

The results from the survey are described in five areas: (1) the rank order for 13 common sources of on-the-job stress; (2) differences in rank order of stressors by job category; (3) differences in rank order of stressors by industry; (4) differences according to relative stressfulness of the overall work environment--whether respondents describe their overall office environment as "pleasant" or as "very stressful"; and (5) the effects of good office air quality versus poor air quality and ventilation conditions. In addition, a very limited amount of information about video display terminal (VDT) users compared with nonusers is given.

In order to introduce the group of workers under discussion, a description of the job and personal characteristics of the survey respondents follows.

I considered 18 job titles (Table 1). The largest are: secretaries of all types, accounting for about 30% of all respondents; clerks and typists (about 20%); and office workers employed in specialized jobs in banking and insurance (about 7%). Specialized clerical jobs in the financial industries are typically very rationalized and often involve the use of computer technology.

TABLE 1
Jobs considered in the survey

| By Major Job Titles [(a)] | % of 960 | No. |
|---|---|---|
| All secretaries | 39.3% | 291 |
| General secretary | 15.0 | 144 |
| Exec./admin. secretary | 8.1 | 78 |
| Legal secretary | 7.2 | 69 |
| Clerks and typists | 22.5 | 216 |
| General clerks | 16.5 | 159 |
| Accounting clerks | 2.5 | 24 |
| Typists | 3.4 | 33 |
| Specialized clerical jobs in banking and insurance | 7.1 | 68 |
| Insurance spec. | 5.7 | 55 |
| Supervisors | 4.6 | 44 |
| Data processing workers | 2.7 | 26 |

[(a)] 18 job titles were analyzed.

By industry of employment, there are concentrations in: banking and other financial, accounting for 20.4% of all respondents; insurance (18%); and the public sector (19.4%). Law firms employ 9% of the respondents.

The mean age is approximately 29 years old; about one-third of the office workers answering the question on their age are over 30. About 20% of the survey respondents are black and other minority workers. Twenty-three men answered the survey (2.4% of the total).

Of 759 respondents who answered questions on their income, 32.9% are sole supporters of themselves or their families. Of these sole supporters, 77% have dependent children. One in four respondents ia a working mother; one in eight is the mother of two or more children. As for marital status, 36.5% report that they are married; 41.6% are single; and 17.1% are divorced, separated, or widowed; 13.5% of all respondents are divorced. The median family income for those who gave income information is $15,000, with 36% reporting total family income of $12,000 or less. In banking and college campus employment, $12,000 is the median.

The 960 respondents are a relatively young group, with a relatively high proportion of sole supporters, although overall they compare well with Bureau of Labor Statistics data on clericals in Cleveland and Boston (ref. 2, 3).

Rank Order of Stressors

Of all office workers answering questions on the stressfulness of their jobs (915), 27% describe their overall office environment as "very stressful," 49% rate it as "somewhat stressful," while 24% describe their overall work environment as "pleasant." Thirteen major sources of job stress, drawn from a review of the research literature, were listed (refs. 4 through 20). Respondents were asked to check the five greatest sources of stress on their jobs.

"Lack of promotions or raises" and "low pay" emerge as the top two sources of job stress--about half of all respondents cite both of these (Table 2). Two in five office workers cite "monotonous or repetitive work" and more than one-third cite "lack of input into decision-making." Heavy workload, an unsupportive boss, supervision problems, and unclear job descriptions are identified by more than one in four respondents as major sources of stress. More than one-third of all respondents indicate that either "an unsupportive boss" or "supervisory problems" are stressors on their jobs, making supervision problems more prevalent than they initially appear.

TABLE 2
Office Worker Health and Safety Survey (a)

| Rank Order | Sources of stress on the job | Adjusted frequency response b |
|---|---|---|
| 1 | Lack of promotions or raises | 51.7% |
| 2 | Low pay | 49 |
| 3 | Monotonous, repetitive work | 40 |
| 4 | No input into decision-making | 35.1 |
| 5 | Heavy workload/overtime | 31.5 |
| 6 | Supervision problems | 30.6 |
| 7 | Unclear job descriptions | 30.2 |
| 8 | Unsupportive boss | 28.1 |
| 9 | Inability or reluctance to express frustration or anger | 22.8 |
| 10 | Production quotas | 22.4 |
| 11 | Difficulty juggling home/family responsibilities | 12.8 |
| 12 | Inadequate breaks | 12.6 |
| 13 | Sexual harassment | 5.6 |

(a) The survey was distributed in Cleveland and Boston in the fall of 1980.
(b) Adjusted frequencies are based on 915 respondents answering the questions on stress (95.3% of the total 960 survey respondents).

CASE STUDIES FROM THE FOLLOW-UP INTERVIEWS

Case studies from the follow-up interviews help to bring to life the daily experience of stress in clerical work and give an idea of how these common sources of stress interact.

Monotony, repetitive work, and rapid work pace are especially common. "I've been doing this job for 10 years, and I've been tired for 10 years," says a 30-year-old data entry operator for a Cleveland utility company. "It's the monotony that does it. I'd like to know what it feels like not to be tired."

In the same department of the utility company, a co-worker describes recently increased pressures where she works:

> Everything seemed just fine at work until last summer, when a company-hired management consulting firm came into our department, supposedly to study how management could work better. But instead of improving management's operation, the consultant began to carefully measure and time our production speed! We used to have to process a maximum of 4,000 checks a day. Now 4,000 has become the minimum--that's one check every six seconds--and the average they require is between 5,000 and 6,000--about four seconds per check.
>
> Now they are treating us just like machines, expecting that everyone can do exactly the same amount every day, no matter how hard or easy checks are to process. It's hard to keep your temper from flaring when you're so tense about meeting production quotas.

Companies use work performance monitoring to enforce production standards and sometimes to set up what can be called a "floating" rate of pay.

Rose re-entered the work force after 20 years away and was quickly hired as one of 12 CRT operators in a downtown Cleveland publishing company. She found that office work had changed a great deal during her years away from the work force.

> The chairs were good, and the machines were adjustable, too, but I have never been confined to one place doing key entry at such a pace. The computer at one end of the room keeps track of the keystrokes you do. The more keystrokes, the more money you might get. At the end of the day, the figures for all of us were posted. You look at your speed, you look at everyone else's and you say, "Tomorrow I'm going to do better." They get you thinking just like they want you to, you're really pushing hard.
>
> One day....I was sitting there at the terminal, entering data, reading from the screen, and I suddenly thought I was having a terrible nightmare. I didn't know where I was. It took all the discipline I had to sit there quietly, until the whole thing passed.

She was afraid to discuss this incident, afraid she might be losing her mind, until she read about our project in the Cleveland group's newsletter. Rose has since changed jobs and works in a more conventional office.

The strain of extremely fast-paced and computer-controlled work at display screens affects not only employees who once did traditional office work. Rose described a co-worker at the publishing company.

> One girl who is also a CRT operator came to the company right out of high school. She's been running a CRT for close to 10 years, and she's fast as the wind. But it's really affected her personality. I used to wonder if something was wrong--she had no exuberance. One morning she turned to me and said, "Rose, as soon as I sit down at the machine in the morning I feel like I'm going to cry."

Lack of promotions and lack of upward mobility are major problems for women in the office. One bank employee has trained six men who were promoted over her in a 10-year period.

Lack of respect and lack of recognition of abilities and accomplishments are part of the daily experience for clericals and secretaries. Amanda, who works for a major industrial firm in Cincinnati, paused at her VDT momentarily just as her manager walked in. "What are you doing?" he demanded. "I'm just thinking," she replied. "Get back to work," he snapped. "You're not paid to think, I am."

Clerical workers frequently complain that they are "treated like machines" or that they are "treated like children," and that they are rarely given credit for the contributions they make to the overall work effort.

A secretary at an accounting firm in a prestigious downtown office in Cleveland comments on the high turnover:

> I've been here almost a year, and I've got seniority among the secretaries. It's the strain--everything must be done under the deadline. We end up rushing around like crazy to make up for their delays in getting the job done so it can be typed. We are forever getting in early and then staying late in order to use the word processor while no one else is waiting for it.

Another secretary adds,

> You know, when the boss brings new clients through the office to show them around, he'll point right at me working at the word processor and say, "Here we have our wonderful new Lexitron," and then moves right along. He doesn't even bother to introduce me--just the machine!

Both women experience almost daily headaches, nervous stomachs, and shaky hands. They are upset that another one of the secretaries who has high blood pressure was recently sent to the hospital for tests. They feel that their employers don't care: "The place looks gorgeous, and that's where the management's priorities lie. They're not really as interested in efficiency as they are in using people up and pushing them out the back door."

Unsupportive bosses and supervision problems often go hand in hand with inability or reluctance to express anger and lack of input into decision-making. More often than not, women report that there is no means to solve workplace problems, that arbitrary treatment prevails, and that procedures, when they exist, are unfair or ineffective. Women report that in many cases they have been threatened with loss of a raise, loss of a promotion, or the loss of their jobs if they bring up issues.

Linda has worked for a very large brokerage firm for many years, for the past 3 years in Boston. She is secretary to the office manager and when he is out of the office, it is assumed that she will take over. Her boss is very unsupportive. When she asks to see her job description, he says things like "I'll get on that," but he never does. Linda feels she should get a raise because she's responsible for the office running smoothly when he is away. But she's been told by her manager, "If you don't like it here, leave," in just so many words. In fact, the company makes it very clear that there are many applicants for her type of job, and she was told that they could probably fill her vacancy (if she left) that same day!

## SURVEY RESULTS ON WORK/HOME RESPONSIBILITIES

One rather surprising finding in the rank order of sources of stress is the low rate given for "difficulty juggling work schedule with home/family responsibilities." Women indicated in follow-up interviews that they felt they can handle the juggling, and that it is the quality of their work life that creates the most stress for them.

For the 117 women who did include this factor in the top five sources of stress, health problems were more prevalent than for the survey group as a whole; they reported work time lost owing to health problems at a rate 30% higher than the rate for all respondents. Not surprisingly, these women are more likely to have children and to be married, separated, or divorced than all respondents. Yet, they represent no more than 20% of women with children, or of ever-married women respondents.

SEXUAL HARASSMENT

Sexual harassment was reported as a major source of stress by 51 women. Although the least frequently cited source of stress, sexual harassment has particularly severe effects. Among women who experience insomnia, severe exhaustion, and nervous stomach problems simultaneously, one in six (17%) was the victim of sexual harassment compared with only 5.6% of all survey respondents.

DIFFERENCES BY JOB TITLE

There are some differences in the rank ordering of stressors and the prevalence of problems related to job categories (Table 3).

TABLE 3
Sources of stress: differences by occupation

| | General secretaries n=136 | General clerks n=153 | Legal secretaries n=68 | Specialized jobs in insurance and banking n=66 | Data processing n=76 |
|---|---|---|---|---|---|
| Lack of promotions | 29.4% | 64.1% | 41.2% | 67.1% | 73.6% |
| Low pay | 52.94 | 61.4 | 36.8 | 63.64 | 26.9 |
| Monotony | 39.7 | 45.8 | 35.3 | 45.5 | 73.6 |
| No input into decision-making | 31.6 | 34.0 | 36.8 | 47 | 46.2 |
| Unclear job descriptions | 27.9 | 28.82 | -- | 31.8 | 19.2 |
| Heavy workload/ overtime | 27.2 | 22.9 | 39.7 | 33.3 | -- |
| Unable to express anger | 22.8 | 25.5 | 32.4 | 25.8 | -- |
| Unsupportive boss | 19.1 | 39.2 | 26.5 | 22.7 | 53.9 |
| Supervision problems | 17.65 | 37.2 | 26.5 | 27.3 | 53.9 |
| Inadequate breaks | 14.7 | -- | 19.1 | -- | 7.7 |
| Production quotas | -- | 26.1 | -- | 37.9 | 65.4 |
| Work/home responsibilities | -- | -- | 17.6 | -- | 26.9 |

There are three general points that can be made about these occupational differences: (1) certain stressors appear to be universal, cutting across job lines and appearing among the top five causes of job stress in almost every job. Most noteworthy are monotony, lack of promotions, and "no input into decision-making." "Low pay" is among the five most commonly cited stressors in all jobs, with the exception of data processing workers, shown here.

(2) Some stressors vary significantly according to the job category examined. "Heavy workload or overtime" is the #2 problem for legal secretaries, although it ranks fifth for all respondents. For legal secretaries, the problem is overtime _per se_, according to information from the interviews with respondents. "Inadequate breaks," while twelfth for the whole survey group, ranks ninth in importance to legal secretaries, who describe being expected to work through their breaks, to work over lunch, and to be available for overtime with little or no advance notice.

The importance of production quotas as a major cause of stress is evident for data processing workers and for office workers in specialized banking and insurance jobs, ranking third in frequency compared with its position as the tenth ranked problem for all survey respondents.

(3) Other differences occur between job categories in the prevalence of problems. For example, while "lack of promotions" is a universal problem, much higher proportions of clerks, specialized bank and insurance workers, and data processing operators cite this problem compared with the proportion of secretaries and legal secretaries. Generally, higher percentages of office workers in the lower-status jobs cite a given stressor than those in the higher-status jobs.

One note in dealing with job titles in office work is in order. "Unclear job descriptions" are problems for more than one in four survey respondents. It is very important for researchers to determine what people actually do at work, since very often what they are called does not reflect real job duties or responsibilities. A person's job title may under-represent or over-represent what he or she does. Lack of clarity in job descriptions is even worse among unorganized workers.

## DIFFERENCES BY INDUSTRY OF EMPLOYMENT

There are also differences in the rank order of stressors by industry of employment (Table 4). The universal problems are "lack of promotions" and "no input into decision-making." The sharpest difference is for "low pay," which varies in degree and in priority. Close to 60% of respondents in both the banking and finance and the insurance industries cite low pay as a problem, and it ranks first in the list of stressors. In law firms and in public sector employment, low pay ranks third, with 36.9% and 43.0% citing low pay, respectively.

This pattern reflects real differences in pay scales (ref. 2, 21). The median family income for respondents employed in banking is $12,000 in contrast to $15,000 for all the survey group. Nationally, banking and insurance pay from 8% to 19% below national averages for clerical pay by occupation, below already-low average salaries, according to Bureau of Labor Statistics data (ref. 2).

TABLE 4
Sources of stress: differences for selected industries

| | Banks and finance n=184 | Insurance n=163 | Public sector n=179 | Law firms n=84 |
|---|---|---|---|---|
| Low pay | 59.2% | 57.1% | 43.0% | 36.9% |
| Lack of promotions | 54.4 | 56.4 | 52.5 | 39.3 |
| Supervision problems | 33.2 | -- | 40.8 | 26.2 |
| Unclear job description | 33.2 | 27.6 | 29.6 | 16.7 |
| Monotony | 32.6 | 42.3 | 43.6 | 35.7 |
| Heavy workload/overtime | 31.5 | 27.6 | 33.5 | 39.3 |
| No input into decision-making | 31.5 | 41.1 | 38 | 32.1 |
| Unsupportive boss | 30.4 | 25.7 | 36.3 | 20.2 |
| Production quotas | 15.8 | 33.1 | 25.1 | -- |
| Unable to express anger | 21.7 | 35 | 16.2 | 32.1 |
| Inadequate breaks | -- | 28.6 | -- | 20.2 |

"PLEASANT" VERSUS "VERY STRESSFUL" JOBS

As mentioned earlier, about half of the respondents describe their overall office environment as "somewhat stressful," about 27% describe their jobs overall as "very stressful," and 24% rate theirs as "pleasant." When the "pleasant" group and the "very stressful" group are compared, there are highly significant differences for each source of stress listed. For the "very stressful" group, "heavy workload or overtime" ranks much higher, as the third-ranked problem, while "monotony" is lower in rank order, compared with all respondents. "Supervision problems" and "unsupportive bosses" are cited by 45% of those who describe "very stressful" working conditions, ranking fourth and fifth. Among office workers who rate their jobs as "pleasant," only 13.6% cite heavy workload, 11% cite supervision problems, and only 9.6% cite unsupportive bosses as sources of stress.

Differences in reported health problems are highly significant for 11 symptoms listed in Table 5. Eyestrain, headaches, and muscle strain are substantial problems for all respondents, but the majority of the "very stressful jobs" group have these complaints compared with about one in three of those in the "pleasant" group. There are striking differences for exhaustion or severe fatigue at day's end, insomnia, and digestive/stomach problems.

The relative stressfulness of the overall work environment is significantly correlated with good or bad air quality and ventilation reported by respondents, and with overcrowding, a complaint for 30% of all respondents (ref. 22).

TABLE 5
Job-related health problems

| Health Problem | Incidence | | |
|---|---|---|---|
| | Respondents rating their jobs as: | | |
| | Average n=960 | Pleasant n=219 | Very stressful n=248 |
| Eyestrain | 54.0% | 39.3% | 68.2% |
| Headaches | 51.2 | 30.6 | 67.7 |
| Muscle strain, pain in back or neck or both | 46.5 | 30.6 | 57.7 |
| Exhaustion or severe fatigue at day's end | 39.6 | 20.55 | 64.9 |
| Digestive/stomach problems or heartburn or both | 24.0 | 12.3 | 37.5 |
| Pain in arms, shoulders, or both | 20.7 | 10.05 | 26.6 |
| Problems from constant sitting | 19.6 | 12.3 | 25.0 |
| Insomnia, trouble sleeping | 18.3 | 7.8 | 31.45 |
| Aching wrists or tendons of the hands | 10.7 | 5.5 | 15.3 |
| Nausea or dizziness | 10.6 | 3.65 | 14.5 |
| Skin rashes or irritation | 8.1 | 5.94 | 11.7 |

## OFFICE AIR QUALITY

Problems with office air quality emerge as the complaints most often cited by respondents, along with job stress. More than 80% report temperature extremes (too hot, too cold, or both); more than 70% say that the supply of fresh air in their offices is inadequate or nonexistent--fully 40% say there is "no fresh air at all" where they work.

Two subgroups of respondents were compared. The first group consists of 81 respondents who report all of these "positive air quality" conditions: a good supply of fresh air, good air circulation, no irritating fumes, no problems with temperature, drafts, or dustiness. The second group consists of 153 respondents who report problems in each of the categories mentioned.

TABLE 6
Correlation between air quality conditions and stressfulness

| | % describing their jobs as: Very stressful | Pleasant |
|---|---|---|
| Positive air quality and ventilation conditions | 14.8% | 40.0% |
| Poor air quality and ventilation conditions | 33.0 | 20.0 |
| | Lost time from health problems | |
| Positive air quality and ventilation conditions | 7.4% | |
| Poor air quality and ventilation conditions | 36.0 | |
| All respondents | 22.6 | |

Table 6 illustrates the correlation between good or poor air quality and relative stressfulness described by respondents. One-third (33%) of respondents reporting air quality problems also describe their jobs as "very stressful," while only 14.8% of those who report good air quality describe "very stressful" work environments. Conversely, two in five respondents in the "positive air quality" group describe their jobs overall as "pleasant," compared with only one in five of those reporting air quality problems.

Among all survey respondents, 22.6% have lost time from work as a result of health problems. More than one in three respondents in the "poor air quality" group have lost time from work, while only 7.4% of those in the "positive air quality" group have lost time owing to health problems. The respondents describing air quality problems also report much higher rates of health complaints for 11 symptoms, sometimes at rates twice or three times the incidence for the "positive air quality" group.

Secretaries are overrepresented in the "positive air quality" group, while clerks are overrepresented among those in the "poor air quality" group. When survey results are analyzed for 18 job categories, clerks, data processing operators, customer service representatives, and women in specialized banking and insurance jobs are more likely to work in areas they describe as too hot, dusty, drafty, without enough fresh air, in areas where air circulation is poor. These employees are more likely than others to describe their overall working conditions as "very stressful," and their work environment as "overcrowded."

Following is just one case study related to indoor air quality problems we came across in Project: Health and Safety.

> Jane works for an insurance company in Boston. At least half of the 50 women in her department have developed eye problems since the company moved into the building 2 years ago when it was brand new. Their work is visually demanding as almost everyone works at a VDT all day. Jane has lost the ability to produce tears, and must rely on a tear replacer prescribed by her doctor.
>
> "When the irritation starts to get out of control," she says, "I get allergic conjunctivitis. We think the irritating fumes are from a lack of fresh air, not changing the air filters often enough, and the photocopiers in the room. Smoke hangs in the air like a cloud. Management claims there is nothing wrong with the air, yet everyone who goes to their eye doctor hears the same thing: 'It's from the building'."

But what "it" is, we may never know. Jane is afraid to call in public health officials because, as the only one who has ever gotten up her nerve to raise the issue with the management, her anonymity cannot be guaranteed. She and her co-workers have reason to be afraid. When she talked to her department manager, she was told that her "complaint" would be entered on her work record under a section called "poor attitude," which would affect her annual salary review. In other words, she was told to keep quiet or lose her raise with a permanent mark against her in her personnel file.

VDT USERS

More than one in three of our survey respondents (37.2%) use VDT's, CRT's, or word processors in their jobs. In terms of health complaints, our findings on VDT users confirm numerous other studies. Unfortunately, we cannot say much about office workers using VDT's who answered our survey, because we did not ask what proportion of time per day is spent at a terminal or for other details. Further questions will be included in future surveys.

Two sources of stress are significantly higher in rank order for VDT users compared with nonusers: production quotas and "inadequate rest breaks." Among VDT users, 27% complain of stress-inducing production quotas among the top 5 sources of stress, compared with 18.1% of nonusers; 15.4% of VDT users cite inadequate rest breaks among the 5 greatest causes of stress, compared with 10% of nonusers. In other regards, the two groups are very similar in their rank ordering of stressors.

A much higher proportion of VDT users report overcrowding, and say they get back strain from poorly designed chairs. VDT users generally report higher incidence of environmental problems, but only these two are significantly higher.

Five health symptoms are significantly higher in incidence among VDT users compared with nonusers: eyestrain (60.8% versus 49.4%); back or neck strain (53.2% versus 42.5%); aching wrists (13.7% versus 9%); and skin rashes (12% versus 5.8%). VDT users account for 44% of all respondents who feel that their health has declined as a result of their present jobs, and a significantly higher percentage of VDT users have lost time from work owing to health problems (27%) than nonusers (20%).

Office workers using VDT's are younger as a group, with 60% between the ages of 20 and 30, and nearly 80% under 35 years of age.

CONCLUSION

In conclusion, there are two comments to be made on the survey results, and two points to be raised concerning unorganized women office workers.

On the survey: (1) there are some differences between this overwhelmingly female group and previous stress studies conducted among male workers. The prominence of "low pay" and "lack of promotions" as top sources of stress contrasts with previous studies in which men rank pay farther down in a list of sources of job stress (ref. 4). This finding is often questioned, with the suggestion that "low pay is not really that important." But we believe that this genuinely reflects the different wage structure for women workers, related to the different occupational structure of the women's work force. In 1979, 28% of all full-time, full-year women clerical workers earned less than $7,800; 82% earned less than $13,000 according to the U. S. Department of Labor, less than the average annual earnings for all male workers (ref. 2).

(2) Our survey results differ somewhat from studies and surveys conducted among unionized workers. For example, the Communication Workers of America (CWA) does not find low pay and lack of promotions at the top of workers' lists of job stressors, since these issues have been handled through collective bargaining for many decades. Unionized clericals average 30% higher wages than nonunionized clerical workers (ref. 23). On the other hand, "no input into decision-making" emerges again and again as a source of job stress among all types of workers, as does "monotonous or repetitive work."

The last two comments are related to the situation of unorganized workers who have no union or staff association to represent them. First, high turnover is unfortunately a policy rather than a problem in certain jobs and industries, particularly the lower-status clerical positions and some key-entry jobs. This practice is most common in the low-paying and highly automated finance industries, which employ very large numbers of office clericals. The deliberate maintenance of high turnover rates is facilitated by the introduction of new office technology designed to reduce training time and make employees more interchangeable. The economic incentives for management in doing this are: that employees do not accrue seniority, retirement benefits, or substantially higher salaries based on incremental raises, and that the work force does not become stable enough to organize in its own behalf. Social cohesion is also undermined among clericals, reducing the potential for social support from co-workers.

Secondly, unorganized clerical workers face serious obstacles when they seek to address health issues, despite protections contained in Federal laws. It is very important for researchers in occupational health to be sensitive to the lack of job security and the often arbitrary environments many women office workers face.

The level of interest in occupational health is extraordinarily high among women clerical and secretarial workers themselves. The Working Women Education Fund together with Working Women will continue our project on clerical workers' health and safety in Boston and Cleveland, and extend it to two more cities in 1981-82. The revised Office Worker Health and Safety Survey will be distributed in the two new cities, and we also plan to question Boston and Cleveland office workers with a more detailed survey on job stress.

REFERENCES


1 Warning: Health Hazards for Office Workers, Working Women Fund, Cleveland, Ohio, 1981.

2 U. S. Department of Labor, Bureau of Labor Statistics, Area wage surveys for selected metropolitan areas industry wage surveys; National professional, administrative, technical and clerical pay, Washington, D. C., 1980.

3 U. S. Department of Commerce, Bureau of the Census, Current population reports, special studies, series P-23, No. 107, Families Maintained by Female Householders, 1970-79, Washington, D. C., 1980.

4 R. D. Caplan, S. Cobb, P. French, Jr., R. V. Harrison, S. R. Pinneau, Jr., Job demands and worker health, main effects and occupational differences, D.H.E.W. (NIOSH) Publication No. 75-160, U. S. Department of Health, Education and Welfare, PHS, CDC, NIOSH, Cincinnati, Ohio, 1975.

5 G. L. Cooper and J. Marshall, Occupational sources of stress: a review of the literature relating to CHD and mental ill health, Journal of Occupational Psychology, 49, 1976. Reprinted in U.S.D.H.E.W., Proceedings of Conference on Occupational Stress, NIOSH Publication No. 78-256, 1978.

6 G. L. Cooper and R. Payne, Stress at Work. John Wiley & Sons, New York, N. Y., 1978.

7 M. J. Dainoff, Occupational Stress Factors in Secretarial/Clerical Work: Annotated Research Bibliography and Analytical Review, U.S.D.H.E.W., NIOSH, Cincinnati, Ohio, 1979.

8 M. Frankenhaeuser and B. Gardell, Underload and overload in working life: outline of a multidisciplinary approach, Journal of Human Stress, 2, 1976, pp. 35-46.

9 S. G. Haynes and M. Feinleib, Women, work and coronary heart disease: prospective findings from the Framingham Heart Study, American Journal of Public Health, 70, 1980.

10 R. A. Karasek, Jr., Job demands, job decision latitude and mental strain: implications for job redesign, Administrative Science Quarterly, 24, 1979, pp. 285-308.

11 Y. Komoike and S. Horiguchi, Fatigue assessment on key punch operators, typists & others, Ergonomics, 14, 1971, pp. 101-109.

12 E. C. Moore, Women and Health U. S. 1980, Public Health Reports, supplement to the September/October 1980 issue, U. S. Public Health Service, Washington, D. C., 1980.

13 E. Palmore, Predicting Longevity: a follow-up controlling for age, Gerontologist, 9, 1969, pp. 247-250.

14 S. M. Sales and J. House, Job dissatisfaction as a possible risk factor in coronary heart disease, Journal of Chronic Diseases, 23, 1971, pp. 861-873.

15 M. J. Smith, M. J. Colligan, J. Hurrell, Jr., A review of NIOSH psychological stress research, in Proceedings of Conference on Occupational Stress, NIOSH Publication 78-256, 1977.

16 M. J. Smith, Recognizing and Coping with Job Stress, NIOSH, Cincinnati, Ohio.

17 J. M. Stellman, Women's Work, Women's Health: Myths and Realities, Pantheon, New York, N.Y., 1977.

18 U. S. Department of Health, Education and Welfare (D.H.E.W.), PHS, CDC, NIOSH, Cincinnati, Ohio, Proceedings of Conference on Occupational Stress, NIOSH Publication No. 78-156, March 1978.

19 U. S. Department of Health, Education and Welfare (D.H.E.W.), Reducing Occupational Stress, Proceedings of a Conference, NIOSH Publication No. 78-140, April 1978.

20 U. S. Department of Health, Education and Welfare (D.H.E.W.), Occupational Health and Safety Symposia 1977, NIOSH Publication No. 78-169.

21 Office Salaries Directory (Administrative Management Society), Willow Grove, Pennsylvania, 1979-80.

22 J. Gregory, Office Air Quality, Tight Buildings and Job Stress--the Impact on Women Office Workers, paper presented at the American Industrial Hygiene Association/American Conference of Government Industrial Hygienists Annual Conference, Portland, Oregon, 1981.

23 U. S. Department of Labor, Bureau of Labor Statistics, Earnings and Other Characteristics of Organized Workers, Bulletin 2105, Washington, D. C., 1980.

# PHYSICAL HEALTH OF CLERICAL WORKERS

LOIS M. VERBRUGGE
Institute for Social Research
P. O. Box 1248
The University of Michigan
Ann Arbor, Michigan 48106

## ABSTRACT

The Framingham Heart Study has shown high rates of coronary heart disease among clerical women, compared with other working women and housewives. This paper asks if American clerical women also have a poor health profile. In fact, they have the best overall profile compared with other occupation groups. Clerical women have low rates of injury and chronic limitation, little restricted activity, and moderate health services use. Different results for Framingham and the U. S. suggest a cohort effect: the Framingham clerical women worked during the 1940's to 1960's, and they experienced great pressures from both their jobs and family lives. These pressures are now less concentrated in clerical jobs, and they may actually be lower than in prior decades. In sharp contrast to the women, American clerical men have a poor health profile. They are especially likely to have job limitations from chronic conditions and to have recent hospitalization. This suggests that some men switch to clerical jobs when they develop severe health problems. Data for these national profiles come from the National Health Interview Survey. A Detroit health survey is also examined, and it buttresses the hypotheses that (1) unsatisfying roles are a risk factor for poor health (a social causation hypothesis), and (2) men with health problems gravitate toward clerical jobs (a social selection hypothesis). We suggest that people who experience combinations of risk factors (from job, family, lifestyle, psyche) become especially vulnerable to illness and injury. Role burdens, role dissatisfactions, and weak coping skills are especially important to identify for employed women, whether they are clerical workers or in other occupations.

## INTRODUCTION

Current concern about the health status of clerical workers comes from several sources. First, the introduction of high technology machines into offices is creating new tasks and task scheduling for clerical workers (ref. 1). Occupational health specialists want to determine if the machines and work patterns pose new occupational hazards for clericals. Second, clerical jobs are the most common occupation for American women, involving one-third of currently employed females (ref. 2). Women's health advocates are understandably interested in the well-being of this large group (ref. 3, 4).

Third, the Framingham Heart Study has shown that clerical women have higher incidence rates of coronary heart disease (CHD) than other working women or housewives (ref. 5). Health scientists and planners want to know if this poor health outcome is true for American clerical women generally and if so, why. Overall, the conjunction of new office milieux, a large population-at-risk, and the Framingham results are urging more research on clerical women's health.

By contrast, there is little current interest in clerical men. They too face new office environments. But clerical jobs are uncommon for American men, involving only 5% of employed males. In addition, the Framingham study found that clerical men had lower CHD incidence rates than men in other occupations. (In fact, their rates were even lower than clerical women's--an atypical result, since other white-collar men and blue-collar men had notably higher rates than their women counterparts.) Here, the small population-at-risk and the favorable Framingham results explain why clerical men have not been singled out for attention.

This paper has two main purposes: first, to discuss the health profile of clerical workers in the United States and also Detroit; and second, to look at the role burdens and role satisfactions that clerical workers experience. The paper is organized as follows: we first examine health differentials by employment status (currently employed versus nonemployed) and by job class (white-collar, blue-collar, service, farm). Then we focus on one occupational group--clericals--and ask if women and men clericals have unusually poor, or good, health profiles compared with other workers. We discuss risk factors that lie behind poor health, then look at one set of them--role burdens and satisfactions--and ask if clerical workers are unusually pressured or unhappy compared with other workers. Finally, we compare the national results with Framingham, and we suggest that Framingham clerical women (who worked mainly in the 1940's to 1960's) confronted greater risks than contemporary clerical women.

## DATA SOURCES

Data came from the National Health Interview Survey and from a Detroit health survey.

The National Health Interview Survey (NHIS) has been conducted since 1957 by the National Center for Health Statistics. It provides annual data for the U. S. population, based on a sample of about 80,000 households. The interview focuses on acute conditions, chronic conditions that cause activity limitations, physician and dentist care, hospitalization, and restricted activity (bed days, work loss, other reduced activities) resulting from illness or injury. Rates for age-sex groups are published routinely, but rates by employment characteristics are more occasional. For this paper, we use NHIS data from four periods: 1977-78, 1975-76, 1972-73, and 1961-63. For 1977-78 and 1972-73, we use tabulations published in Vital and Health Statistics (Ref. 6, 7). The 1977-78 data provide rates by employment status for age-sex groups. The 1972-73 data have rates by job class and by occupation, for age-sex groups. For 1975-76 and 1961-63, there are rates by job class and by occupation, for each sex (but not for age-sex groups). A further note: the 1975-76 rates refer to currently employed people, but the 1972-73 and 1961-63 rates refer to labor force participants (see definition in next section). For more details about the NHIS sample and questionnaire, see ref. 7.

We also use data from the Health in Detroit study, which surveyed a representative sample of white adults residing in the Detroit metropolitan area in 1978. The study had three stages: an initial interview covered aspects of current and recent health; many of its questions were similar to NHIS items. Respondents then kept Daily Health Records for 6 weeks. Each day, they gave information about their overall health status, specific physical symptoms, curative actions, and preventive actions. At the end of the diary period, a termination interview was conducted, asking respondents about their reactions to the diary task and their perceptions of health changes during the 6 weeks. Altogether, 714 people (412 women, 302 men) had an initial interview, and 589 (346 women, 243 men) kept one or more weeks of Daily Health Records. For this analysis, we use health information from both the initial interview and the health diaries. Information about roles comes from the initial interview. For more details about the survey design and response rates, see ref. 8, 9, 10.

## DEFINITIONS

Employment characteristics are defined as follows: employment status indicates whether people have a job at the time of interview. Three categories are used: currently employed, unemployed (not working but looking for a job), and not in labor force (not working and not looking for a job). The labor force consists of currently employed + unemployed people. The nonemployed population consists of unemployed + not in labor force people. Job class refers to traditional socioeconomic groups; white-collar, blue-collar, service, farm. Sometimes, a further split of "upper" and "lower" is used among white-collar workers, and also among blue-collar workers. Occupation indicates the kind of work a person performs. The Census Bureau's classification scheme has 12 major groups. The group "clerical and kindred workers" includes such jobs as bank teller, cashier, file clerk, mail carrier, secretary, stock clerk, telephone operator, and ticket agent. One other note: for currently employed people, job class and occupation refer to their current job; for unemployed people, to their most recent job.

In this paper, health is a very inclusive term, encompassing all indicators of health status and health behavior. Measures of health status are rates of illness, symptoms, and injury, and also subjective ratings of health. Measures of health behavior are restricted activity (short-term disability), chronic limitations (long-term disability for chronic conditions), health services use, medical drug use, and other preventive and curative actions.

## HEALTH DIFFERENTIALS BY EMPLOYMENT STATUS AND JOB CLASS

### Employment status

Table 1 shows that currently employed people have better health than nonemployed groups. Unemployed people rank second, and people outside the labor force rank lowest. These statements hold true for both sexes.

There are several possible reasons for the good health of currently employed people. First, employment increases a person's social and financial resources, allows expression of skills and knowledge, and is an important source of self-esteem. These aspects may protect or even enhance health, by reducing risks of illness and injury and by improving the quality of health care received. Second, employed people may have different health perceptions and attitudes than nonemployed people. Work commitments take up time and mental energy; as a result, employed people may be less aware of symptoms and also less likely to take curative actions for the ones they perceive. Third, health can influence employment status itself. Poor health forces some people to quit work, and it also reduces nonemployed people's chances of finding jobs. In sum, the connection between employment and good health is due to both social causation (how employment affects health) and to social selection (how health influences employment), but their relative importance is not known now.

These employment differentials have been found repeatedly in national and community health studies, including the Health in Detroit study. (See ref. 11, 12, 13, 14, 15, 16, 17).

### Job class

Table 2 shows that job classes differ markedly in their health. We will compare job classes within each sex (for example, white-collar women with the other job classes of women).

TABLE 1
Health by employment status and sex
(1977-78 National Health Interview Survey. Persons ages 17-64)

| | Women | | | Men | | |
|---|---|---|---|---|---|---|
| | Currently employed | Unemployed | Not in labor force | Currently Employed | Unemployed | Not in labor Force |
| Percent who say their health is:[a] | | | | | | |
| Poor | 0.9 | 1.8 | 5.5 | 1.0 | 2.9 | 16.1 |
| Excellent or good | 90.1 | 86.3 | 79.5 | 91.5 | 85.0 | 66.1 |
| Incidence of acute conditions (per 100 persons per year)[b] | 227 | 216 | 197 | 174 | 157 | 171 |
| Respiratory conditions | 117 | 110 | 99 | 88 | 83 | 94 |
| Injuries | 32 | 27 | 24 | 41 | 35 | 32 |
| Percent with activity limitation due to chronic condition[c] | 7.6 | 11.4 | 19.2 | 9.6 | 16.0 | 41.7 |
| Unable to do major activity | 0.1 | 0.4 | 2.6 | 0.3 | 3.6 | 32.1 |
| Limited in kind or amount of major activity | 4.7 | 7.4 | 12.1 | 5.5 | 8.3 | 6.3 |
| Limited in secondary activity only | 2.8 | 3.6 | 4.5 | 3.8 | 4.1 | 3.2 |
| Restricted activity days (per person per year)[d] | 14.1 | 26.9 | 27.0 | 11.1 | 32.1 | 40.9 |
| Bed disability days | 5.4 | 11.0 | 10.5 | 3.4 | 10.2 | 14.8 |

Source: Unpublished tabulations provided by National Center for Health Statistics.

a Response categories are excellent, good, fair, poor.
b These are short-term conditions that caused restricted activity or medical care.
c Adults are asked about their usual activity in the past year: working (a paid job), keeping house, retired, or something else. If working, retired, or something else: current work limitations are asked. If keeping house: current housework limitations are asked.
d Days people cut down their usual activities because of illness or injury.

TABLE 2
Health for job classes (currently employed persons and nonemployed persons) by sex.
(1975-76 National Health Interview Survey, and the Health in Detroit study. Persons ages 17+ for NHIS, and 18+ for Detroit)[a]

| 1975-76 National Health Interview Survey | Women | | | | Men | | | | Women | Men |
|---|---|---|---|---|---|---|---|---|---|---|
| | White-collar | Blue-collar | Service | Farm[b] | White-collar | Blue-collar | Service | Farm | Unemployed | Unemployed |
| Incidence of acute conditions (per 100 persons per year) | 221 | 188 | 217 | 181 | 170 | 173 | 198 | 114 | 220 | 137 |
| Respiratory | 123 | 94 | 111 | 74 | 102 | 87 | 106 | 58 | 111 | 68 |
| Injuries | 28 | 30 | 31 | 36 | 26 | 43 | 35 | 34 | 29 | 36 |
| Injuries at work | 5 | 10 | 6 | 12 | 5 | 23 | 13 | 21 | -- | -- |
| Percent with activity limitation due to chronic condition | 7.9 | 9.2 | 11.1 | 11.7 | 10.9 | 10.3 | 12.8 | 17.2 | 13.1 | 16.0 |
| Limited in major activity[c] | 4.6 | 6.1 | 7.8 | 7.9 | 6.1 | 6.5 | 9.0 | 13.7 | 9.2 | 11.7 |
| Limited in secondary activity only | 3.3 | 3.1 | 3.3 | 3.8 | 4.8 | 3.8 | 3.8 | 3.6 | 3.9 | 4.4 |
| Restricted activity days (per person per year) | 12.8 | 16.9 | 14.4 | 9.7 | 9.9 | 12.2 | 13.9 | 11.0 | 27.6 | 29.3 |
| Bed disability days | 5.1 | 5.8 | 5.5 | 2.4 | 3.1 | 3.7 | 4.7 | 3.0 | 12.3 | 9.6 |
| Work loss days | 5.0 | 7.9 | 6.0 | 3.1 | 3.8 | 6.0 | 6.7 | 4.4 | -- | -- |
| Percent with short-stay hospital episode in past year | 11.0 | 13.2 | 11.6 | 7.9 | 7.4 | 7.5 | 7.8 | 7.5 | 17.3 | 11.3 |
| Average length of stay (days) | 5.9 | 5.9 | 6.3 | 5.3 | 7.2 | 7.5 | 8.6 | 6.8 | 7.6 | 8.7 |
| Percent with physician visit in past year | 84 | 78 | 81 | 73 | 70 | 65 | 70 | 59 | 83 | 65 |
| No. physician visits per year (average for all persons) | 5.7 | 5.4 | 5.3 | 4.5 | 3.9 | 3.4 | 4.1 | 3.3 | 7.0 | 4.4 |
| Percent with dentist visit in past year | 65 | 43 | 47 | 45 | 60 | 42 | 44 | 39 | 53 | 43 |
| No. dentist visits per year (average for all persons) | 2.2 | 1.4 | 1.6 | 0.9 | 1.9 | 1.3 | 1.4 | 1.1 | 1.9 | 1.7 |
| Percent with hospital insurance | 90 | 86 | 75 | 60 | 91 | 84 | 80 | 65 | 62 | 51 |
| Percent with surgical insurance | 88 | 84 | 72 | 59 | 89 | 83 | 78 | 63 | 60 | 50 |
| Out-of-pocket health expenses, including insurance premium($) | 307 | 282 | 252 | 255 | 270 | 201 | 202 | 267 | 247 | 190 |

TABLE 2 (Continued)

| Health in Detroit survey (age-controlled)[d, e] | Women | | | | Men | | | | Women | | Men | |
|---|---|---|---|---|---|---|---|---|---|---|---|---|
| | Upper white-collar | Lower white-collar | Upper blue-collar | Lower blue-collar | Upper white-collar | Lower white-collar | Upper blue-collar | Lower blue-collar | Unemp. | Not in labor force | Unemp. | Not in labor force |
| *n* | 70 | 119 | 32 | 45 | 99 | 47 | 118 | 32 | 27 | 174 | 8 | 38 |
| Self-rated health status (1 = excellent, 5 = poor) | 1.7 | 1.8 | 1.7 | 2.2 | 1.6 | 1.8 | 2.0 | 1.8 | 2.2 | 2.3 | 2.6 | 1.8 |
| Physical feeling during diary period (average; 1 = terrible, 10 - wonderful)[f] | 7.2 | 7.0 | 7.6 | 6.8 | 7.4 | 7.4 | 7.3 | 7.3 | 7.0 | 6.8 | 6.0 | 6.7 |
| No. health problems during diary period | 25.8 | 23.9 | 15.3 | 26.0 | 15.1 | 13.7 | 14.0 | 18.8 | 21.6 | 31.8 | 34.6 | 27.2 |
| No. chronic conditions/symptoms in past year | 3.6 | 4.1 | 3.1 | 4.2 | 3.2 | 3.2 | 3.6 | 3.5 | 3.8 | 5.0 | 2.9 | 5.1 |
| Job limitations due to chronic conditions (an index score)[g] | 1.13 | 1.14 | 1.11 | 1.16 | 1.16 | 1.19 | 1.18 | 1.11 | 1.41 | 1.44 | 2.11 | 1.72 |
| No. restricted activity days in past year | 9.7 | 9.8 | 15.0 | 6.2 | 10.6 | 6.4 | 14.0 | 14.5 | 17.5 | 29.2 | 104.6 | 36.4 |
| No. restricted activity days during diary period | 2.8 | 2.6 | 3.3 | 2.3 | 2.1 | 2.5 | 3.7 | 4.4 | 4.7 | 5.1 | 17.6 | 3.7 |
| No. drugs currently being used for chronic conditions | 1.1 | 1.2 | 1.2 | 1.3 | 1.0 | 0.9 | 0.8 | 0.3 | 1.9 | 1.8 | 1.8 | 1.5 |
| No. prescription drugs used during diary period[h] | 9.6 | 15.9 | 16.6 | 24.3 | 14.0 | 15.8 | 13.9 | 4.4 | 25.2 | 28.4 | 3.4 | 33.5 |

Source: For 1975-76 NHIS, ref. 7

TABLE 2 (Continued)

[a] Job classes are based on Census occupation titles as follows: for NHIS, white-collar = professional/technical, administrator/manager, sales, clerical; blue-collar = crafts, operatives except transport, transport equipment operatives, laborers except farm; service = service except private household, private houehold; farm = farmer/farm manager, farm laborer/farm foreman. For Health in Detroit, upper white-collar = professional/technical, administrator/manager; lower white-collar = sales, clerical; upper blue-collar = crafts, operatives except transport, farmer/farm manager (n = 1), service except private household, private household.

[b] Most rates for farm women have high sampling variability (relative standard error over 30%).

[c] Includes unable to do major activity and limited in kind/amount of major activity.

[d] Data are for white adults. Numbers are based on multiple regressions with predictors for age and job class. (A predictor for morbidity was also included when the dependent variable was a health behavior.) Results shown here are therefore free of age and morbidity differences among job classes.

[e] Items labeled "during diary period" come from the Daily Health Records. All other items come from the initial interview.

[f] Each day, respondents answered "How did you feel physically today?"

[g] Individual respondents are scored as follows: 1 = no limitation, 2 = limited in the kind of job, or in the amount of work at a job, 3 = unable to have a paying job.

[h] Each day, respondents wrote the names of pills, medicines, and treatments used. These are summed across the entire diary period. A drug used on several days is counted on each day.

Looking first at 1975-76 NHIS data: (1) white-collar workers have the best health profile, but they are nevertheless active users of health services. They have low rates of injury, acute respiratory infections (men, but not women), and major chronic limitations. White-collar workers have moderate or low rates of restricted activity and hospitalization, but high physician and dentist visit rates (possibly for more active preventive care). They have the widest health insurance coverage, yet they still spend more of their own money on health than other groups do. This profile holds for both women and men. (2) Blue-collar workers have intermediate rates for most health indicators. There are a few exceptions to this: first, blue-collar women have the highest restricted activity and hospitalization rates, compared with other classes of women. This is not matched by high morbidity, so it suggests that blue-collar women's health problems are particularly serious ones. Second, blue-collar men have higher injury rates than other men, probably reflecting hazards of their jobs. Although serious injuries often lead to impairments and limitations, few blue-collar men report chronic limitation. This hints at a social selection effect: blue-collar men who experience serious injury may have trouble performing their jobs, so they leave the labor force or find white-collar jobs that are less demanding physically. (3) Among service workers, women are especially troubled by chronic problems, and men by acute ones. Women service workers have high chronic limitation rates, but rather low health insurance coverage and personal health expenses. It is possible that women with chronic problems gravitate toward service jobs, where accommodations in scheduling and tasks can often be made to meet personal needs. But these jobs tend to be low-paying and to lack fringe benefit programs. Men service workers have the highest rates of health care (restricted activity, hospitalization, physician visits) among men. This may be in response to acute conditions, which are relatively frequent for them. (4) Farm workers have especially high rates of chronic limitation. This partly reflects their older ages, compared with other job classes. Otherwise, farm workers have very low rates of acute problems, restricted activity, health services use, and health insurance coverage. These workers appear to have little need or desire for curative health care. These statements hold for both sexes (women farm workers are not numerous, but the profile still appears for them).

Using 1972-73 NHIS data, we reviewed rates of acute conditions, restricted activity for acute conditions, and chronic limitations. Differentials by job class are consistent with the profiles above. We have therefore chosen to show only the more recent 1975-76 results in Table 2.

Turning to the Health in Detroit survey, four job classes are defined: upper white-collar, lower white-collar, upper blue-collar, and lower blue-collar (includes service). Health tends to decline as job status does (Table 2): (1) upper white-collar workers appear to be the healthiest group. Although they seldom rank best for any specific indicator, they have a consistent profile of low or moderate morbidity and curative behavior. (One exception: upper white-collar women report numerous daily symptoms in the health diaries. This is parallel to the high acute condition rates found among them in NHIS. We have no ready explanation for the high frequency of short-term problems for these women.) (2) Lower white-collar workers are intermediate for most indicators. (3) Upper blue-collar and lower blue-collar workers tend to be less healthy than the white-collar groups. Women in lower blue-collar jobs clearly have the poorest health profile, compared with other women. No group of men stands out with uniformly poorest health.

The national and Detroit results are very compatible; both show that white-collar workers enjoy better health than other job classes. These workers' frequent visits to physicians and dentists probably reflect good access to services and strong desire to use them, rather than great need. It is not so easy to identify the job class with poorest health. For women, service and farm workers (in NHIS) and lower blue-collar workers (in Detroit) stand out for certain problems. For men, blue-collar workers (including service) stand out for injuries and restricted activity, and farm workers for chronic limitations.

These health differentials point toward both social causation and social selection. Some results reflect how job hazards and possibly health attitudes differ for job classes, leading to different health conditions and health care. But it is also true that blue-collar jobs are often physically demanding, so workers who develop health problems leave such jobs, moving toward white-collar and some service jobs or quitting the labor force altogether.

### The main separator

Employment status separates healthy and unhealthy people more than job classes do. Employed people in any job class tend to be healthier than nonemployed people. In other words, although blue-collar workers have a poorer profile than white-collar workers, they are still better off than nonemployed people. This is especially obvious for indicators of serious or chronic health problems.[1]

Some evidence is shown in Table 2; rates for nonemployed people on the right side can be compared with the job classes on the left. Looking first at NHIS: unemployed people have more chronic limitations and more restricted activity days, hospitalization, and physician care in the past year than the job classes. Despite their high use of health services, they have less health insurance coverage and lower personal health expenses, suggesting that they receive lower quality care. (Two exceptions should be noted: men farm workers have similar levels of chronic limitation as unemployed men do. Also, unemployed people have one advantage over most job classes; they have fewer acute respiratory ailments per year.) Looking next at Detroit data: virtually all indicators show more disability, more curative care, and more frequent symptoms for nonemployed people, compared with the job classes. In sum, both surveys indicate that unemployed and not-in-labor-force people suffer from serious and persistently symptomatic health problems.

### Sex differences

Within an employment status or job class, women usually have higher rates of morbidity and health behavior than men do. Only a few exceptions appear: men are more likely to have chronic limitations; they suffer more injuries at work; and their hospital stays are longer than women's. A comprehensive review of sex differentials appears elsewhere (ref. 18). It concludes that women tend to have more short-term health problems (daily symptoms and acute illnesses), whereas men suffer fewer but more life-threatening health problems.

## CLERICAL WORKERS' HEALTH

Do American clerical workers have an especially good, or poor, health profile? We shall describe women clericals and men clericals separately. Each group is compared with all employed persons and with other occupational groups, of the same sex. We rely mainly on 1975-76 NHIS data, and supplement them with NHIS data for earlier years (1972-73, 1961-63) and with Health in Detroit data.

Clerical women

The National Health Interview Survey reveals very good health for American clerical women, compared with all employed women (Table 3): clericals have very low injury rates, especially injuries at work. Chronic limitations are uncommon, and acute respiratory conditions are average. Reflecting their low morbidity, clerical women have few restricted activity days (including few work loss days) compared with other employed women, and their health services use is average or less. They have relatively high health insurance coverage and average out-of-pocket health expenses. Overall, this is a profile of very good health status and infrequent need for curative care, coupled with good insurance coverage in case serious problems occur.

We reviewed the health profiles of all other occupational groups (*n* = 11) and found that none has such a consistently good health profile as clerical women. (Other groups tend to be more variable, showing excellent health on some indicators but poor health on others. For example, professional/technical women have the widest health insurance coverage, but they are bothered relatively often by acute conditions and some types of injury.) Readers might suspect that clericals are younger than other employed women, and that this explains the good profile. But the age distributions of clerical women and all employed women are very similar. (Fifty-three percent of the clericals are ages 17 to 34, compared with 47% for all employed women.)

National data for other years (1972-73, 1961-63) concur with the profile above: compared with other occupational groups, clerical women have the lowest chronic limitation rates, lowest work injury rates, average rates of acute respiratory conditions, next to lowest restricted activity rates, and below average hospitalization rates. For most types of acute conditions, clerical women take fewer restricted activity days per condition than other women; in other words, they cut down their activities less when ill than other women do. This suggests their acute ailments are mild ones.

In sum: American clerical women have the best overall health profile among employed women.

TABLE 3
Health of clerical workers compared with all currently employed persons (1975-76 National Health Interview Survey, and the Health In Detroit Study. (Persons ages 17+ for NHIS, and 18+ for Detroit.)[a]

| 1975-76 National Health Interview Survey | Women | | | Men | | |
|---|---|---|---|---|---|---|
| | All currently employed | Clerical workers | Rank[b] (1=low, 12-high) | All currently employed | Clerical workers | Rank |
| Incidence of acute conditions (per 100 persons per year) | 215 | 206 | 8 | 171 | 155 | 6 |
| Respiratory | 116 | 117 | 6 | 94 | 88 | 9 |
| Injuries | 29 | 22 | 2 | 35 | 21 | 2 |
| Moving motor vehicle | 3 | 3 [c] | 9 | 3 | 1 [c] | 3 |
| At work | 6 | 3 | 2 | 15 | 6 | 3 |
| At home | 10 | 9 | 4 | 8 | 8 | 7 |
| Other | 10 | 7 | 5 | 11 | 7 | 3 |
| Percent with activity limitation due to chronic condition | 8.8 | 7.4 | 2 | 11.1 | 12.2 | 9 |
| Limited in major activity | 5.5 | 4.4 | 3 | 6.9 | 8.2 | 9 |
| Limited in secondary activity only | 3.3 | 3.0 | 2 | 4.2 | 4.0 | 7 |
| Restricted activity days (per person per year) | 13.7 | 12.2 | 3 | 11.3 | 11.1 | 6 |
| Bed disability days | 5.3 | 4.9 | 3 | 3.5 | 3.5 | 7 |
| Work loss days | 5.6 | 5.0 | 5 | 5.0 | 5.2 | 6 |
| Percent with short-stay hospital episode in past year | 11.4 | 10.9 | 5 | 7.4 | 8.2 | 11 |
| Discharges per 100 persons per year | 13.6 | 13.4 | 7 | 9.3 | 9.9 | 9 |
| Average length of stay (days) | 6.0 | 5.8 | 6 | 7.4 | 8.2 | 11 |
| Percent with physician visit in past year | 82 | 83 | 9 | 68 | 69 | 9 |
| No. physician visits per year (average, all persons) | 5.6 | 5.5 | 9 | 3.6 | 3.9 | 8 |
| Percent with dentist visit in past year | 58 | 63 | 11 | 50 | 52 | 9 |
| No. dentist visits per year (average, all persons) | 2.0 | 2.1 | 9 | 1.5 | 1.5 | 8 |
| Percent with hospital insurance | 86 | 89 | 10 | 86 | 91 | 11 |
| Percent with surgical insurance | 84 | 88 | 10 | 84 | 89 | 11 |
| Out-of-pocket health expenses, incl. insurance premium ($) | 292 | 288 | 7 | 234 | 220 | 6 |
| Hospital | 27 | 24 | 6 | 20 | 12 | 2 |
| Doctor | 74 | 72 | 8 | 53 | 48 | 7 |
| Dental | 58 | 65 | 9 | 48 | 47 | 8 |
| Prescription medicines | 37 | 36 | 5 | 26 | 28 | 8 |

| TABLE 3 (Continued) Health in Detroit Study (age-controlled)[d,e] | Women | | | Men | | |
|---|---|---|---|---|---|---|
| | All currently employed | Clerical workers | Rank[b] (1=low, 12=high) | All currently employed | Clerical workers | Rank |
| n | 202 | 76 | | 254 | 20 | |
| Self-rated health status (1 = excellent, 5 = poor) | 1.83 | 1.85 | 4 | 1.78 | 1.95 | 5 |
| Physical feeling during diary period (average; 1 = terrible, 10 = wonderful) | 7.10 | 7.02 | 4 | 7.36 | 7.46 | 5 |
| No. health problems during diary period | 23.9 | 25.0 | 4 | 14.9 | 10.9 | 1 |
| Very serious, or somewhat serious[f] | 8.0 | 8.9 | 5 | 3.0 | 1.6 | 2 |
| Not very serious | 15.4 | 15.5 | 3 | 11.7 | 9.0 | 1 |
| No. chronic conditions/ symptoms in past year | 3.9 | 4.3 | 5 | 3.4 | 3.3 | 4 |
| Job limitations due to chronic conditions (an index score) | 1.14 | 1.17 | 6 | 1.17 | 1.26 | 6 |
| No. restricted activity days in past year | 9.8 | 9.1 | 3 | 11.6 | 4.7 | 1 |
| No. restricted activity days during diary period | 2.7 | 2.6 | 4 | 3.0 | 1.4 | 1 |
| No. drugs currently being used for chronic conditions | 1.3 | 1.2 | 3 | 0.8 | 1.0 | 4 |
| No. prescription drugs used during diary period | 15.4 | 17.3 | 5 | 13.1 | 15.5 | 5 |
| No. physician visits for curative care in past year | 2.2 | 2.2 | 4 | 2.0 | 3.1 | 6 |
| No. physician visits for preventive care in past year | 1.3 | 1.4 | 5 | 1.1 | 1.2 | 4 |
| No. days with curative medical or dental care during diary period | 0.6 | 0.7 | 6 | 0.7 | 0.5 | 2 |

Source: For 1975-76 NHIS, ref. 7.

a See Table 1 footnotes for details about the health indicators.

b For NHIS, this is the rank among 12 occupational groups. For Health in Detroit, it is the rank among 6 occupational groups (professional/ technical; administrator/manager; clerical; sales; crafts, operatives except transport, farmer/farm manager; nonfarm and farm laborers, all service, transport equipment operatives).

c Most rates for specific injury types have high sampling variability.

d Data are for white adults. Numbers are based on multiple regressions with predictors for age and occupational group. (A predictor for morbidity was also included when the dependent variable was a health behavior.) Results shown here are therefore free of age and morbidity differences among occupational groups.

e Items labeled "during diary period" come from the Daily Health Records. All other items come from the initial interview.

f Respondents rated their problems as very serious, somewhat serious, or not very serious. The categories do not add to the total because of missing information about seriousness.

Clerical women in Detroit do not show this same health advantage (Table 3): compared with all employed women, clericals report more chronic health problems, the most job limitations, slightly more daily symptoms (especially "serious" ones), and more prescription drug use during the diary period. They are average in other respects (self-rated health, restricted activity, medical care). Overall, clerical women in Detroit have a poorer health profile than employed women generally. Because the national and Detroit profiles differ so much, we believe that Detroit clerical women are not representative of American clerical women.

Clerical men

American clerical men have an intriguing profile. The 1975-76 NHIS data show that, compared with all employed men, clerical men have distinctly more chronic limitations and more recent hospitalization (Table 3). Other kinds of curative care are above average or average. Clerical men are advantaged in two respects: they have strikingly low injury rates, especially work injuries.[2] And they enjoy high hospital insurance coverage, which keeps their out-of-pocket hospital expenses low.

Compared with other occupational groups ($n$ = 11), clerical men clearly have the worst health profile among white-collar occupations. Several blue-collar groups (service workers, private household workers) and farmers do suffer more chronic limitations. There is some evidence that the blue-collar men and farmers are troubled more by orthopedic impairments, whereas the clerical men tend to have serious chronic diseases: first, in the 1961-63 NHIS, prevalence rates for limitations resulting from orthopedic impairments are given; clerical men have below average rates, blue-collar groups are above average, and farmers have the highest rates. Second, in the Detroit survey, clerical men rank highest for experiencing a heart attack in the past year and for ever having a heart attack. Their prevalence rates of impairments and of musculoskeletal ailments are average; blue-collar occupations typically have highest rates for these.[3]

National data for earlier years (1972-73, 1961-63) have concurring evidence: clerical men are especially likely to report troubles in the kind or amount of work they do.[4] This is clearest for men ages 45 to 64. Problems doing secondary activities (such as errands and recreation) are average or above average, and hospitalization rates are also above average.

In sum: American clerical men have a poor health profile compared with other white-collar occupations. They are especially likely to have chronic problems that affect the kind or amount of work they can do, and have caused recent hospitalization. The results hint at social selection--that some men with serious health problems move into clerical jobs, which often have lower physical and emotional demands than other jobs.

Clerical men in Detroit repeat the national profile (Table 3): they report more job limitations resulting from health than any other occupation, and they have high rates of medical care and medical drug use in the past year. They do not report much restricted activity in the past year; national data showed a similar result. Clerical men rate their general health status relatively low; only crafts workers rank lower. (This profile of poor health comes from the initial interview, which has a similar time frame as NHIS. The health diaries look different: on a day-to-day basis, clerical men do not have many symptoms or restricted activity, and they say they feel quite good. They do, however, use more drugs than other men do.)

## ROLE BURDENS AND ROLE SATISFACTIONS OF CLERICAL WORKERS

Occupational health specialists look for risk factors of jobs--in the tasks people perform, the physical environment of the worksite, relationships with employers and co-workers, and stresses from task scheduling. In understanding an occupational group's health, job characteristics are certainly important. But we also need to look at other aspects of the group--their other roles, their attitudes about roles, their psychological strengths--to see if they experience other special risks. Health scientists are now actively studying how major life changes (called stressful life events), social support, role burdens, role satisfactions, and coping skills affect physical and mental health. These factors may influence health regardless of a person's occupation. But their significance for occupational health research is how they might interact with job factors to markedly increase health risks. For example, people who become divorced may be more sensitive to job stresses. Or people who feel overloaded by their family and job responsibilities may become fatigued and very vulnerable to health problems. (In statistical terms, these are called interaction effects. A job factor A may pose some risk by itself, but when a nonjob factor B is also present, A's effect increases notably.)

Women's health researchers are especially interested in role burdens and role satisfactions. Women usually spend more time doing household tasks than men do, and they are typically responsible for household management (planning for child care, meals, shopping, etc.). These responsibilities do not diminish much when women are employed. Thus, a job increases women's time constraints and duties. By contrast, a man's main responsibility is his job; his domestic role is secondary and it does not change much when his spouse is employed. In sum, multiple roles (job plus family responsibilities) are more likely to pose stresses for women, and they can affect health negatively in the short run and the long run. (We must also recognize positive aspects of multiple roles for women. Having both job and family attachments offers resources and satisfactions from several sources, rather than just one, and this may confer health benefits.) In addition, satisfaction with one's roles may be even more important for health than the number of roles. People who are content with their responsibilities and activities--however numerous or few--many enjoy health benefits.

Research shows that women with multiple roles actually have better physical and mental health than less-involved women (ref. 15, 19, 20, 21, 22, 23). The same is true for men. In addition, people who are satisfied with their roles have better health than dissatisfied people (ref. 24). These results are only a broad stroke, and scientists must go farther to pinpoint the aspects of multiple roles that affect health (positively or negatively) and to identify workers who have the most difficulty combining job and family roles.

So far, research on clerical workers has focused on job risks, such as sedentary tasks, poor lighting, cramped seating, limited opportunities for promotion, and stress from frequent interruptions and variable workloads. In this paper, we look more generally at clerical workers' roles, and ask if clericals are more pressured or unhappy in their roles than other workers. National data are not available, but the Detroit survey provides some information about job schedules, domestic responsibilities, and role satisfaction.

## Clerical women

Clerical women have many advantages resulting from regular job schedules and limited domestic responsibilities, but they are less happy with their roles and lives than other employed women.

Most clerical women in Detroit work during business hours on weekdays, and do not work on weekends or overtime (Table 4). Compared with all employed women, they have fewer domestic ties; smaller percents are married or have children at home. (Clerical women are not younger; their age distribution is similar to all employed women.) Adding up the number of major roles they have, clerical women are less involved than average. Their time pressures are about average, whether we look at objective time constraints or their subjective feelings. Their responsibility for household income is average or below average. The most striking characteristic of clerical women is their dissatisfaction with job and family roles. Of all occupational groups, they rank lowest or next to lowest for job satisfaction, housework satisfaction, and life satisfaction. (More clericals say they dislike a role, have mixed feelings, or qualify their likes.)

TABLE 4
Job and family characteristics of clerical workers
(Health in Detroit survey)

| | Women | | Men | |
|---|---|---|---|---|
| | All currently employed | Clerical workers | All currently employed | Clerical workers |
| | 210 | 76 | 256 | 20 |
| | (Percents except where otherwise noted) | | | |
| Job characteristics | | | | |
| Work same hours each day | 80.8 | 86.8 | 79.5 | 75.0 |
| Start work between 8 a.m.-noon | 65.0 | 80.0 | 34.3 | 33.3 |
| Stop work between 4-8 p.m. | 61.3 | 78.5 | 55.2 | 73.3 |
| Workday length is 8-9 hrs. | 66.3 | 80.0 | 64.2 | 66.7 |
| Work same days each week | 82.3 | 85.5 | 90.2 | 95.0 |
| Work all 5 weekdays | 77.2 | 87.7 | 89.5 | 78.9 |
| Work on weekends | 25.8 | 7.7 | 35.2 | 36.9 |
| Work 50 or more hours per week | 8.3 | 1.3 | 34.3 | 20.0 |
| Family characteristics | | | | |
| Currently married | 52.9 | 49.4 | 72.3 | 70.0 |
| Parent (one or more own children in household) | 43.8 | 37.7 | 49.6 | 35.0 |
| Own preschool child(ren) in household | 14.2 | 7.8 | 25.2 | 20.0 |
| Index of own child dependency (Based on number and ages of own children in household) | | | | |
| Low dependency | 55.1 | 62.3 | 49.8 | 63.2 |
| High dependency | 13.6 | 11.7 | 18.4 | 10.5 |
| Index of total dependency (Based on own children and elderly adults in household) | | | | |
| Low dependency | 52.6 | 60.5 | 47.0 | 57.9 |
| High dependency | 13.8 | 11.8 | 18.4 | 10.5 |
| Role involvements and responsibilities | | | | |
| Number of major roles (Based on employment, marital, and parent statuses)(average) | 2.0 | 1.9 | 2.2 | 2.0 |
| Number of major activities (Based on employment, marriage, parenthood, volunteer work, other regular activities)(average) | 2.7 | 2.6 | 3.0 | 3.0 |
| Index of role responsiblity (Based on employment, marriage, own child dependency, elderly dependency) | | | | |
| Very high | 13.8 | 9.2 | 35.3 | 26.3 |
| High | 49.7 | 59.2 | 38.2 | 47.4 |
| Percent of household income earned by respondent | | | | |
| Low (1-19%) | 26.9 | 21.7 | 2.5 | 5.3 |
| Modest (20-49%) | 34.6 | 42.0 | 7.9 | 10.5 |
| All (100%) | 27.5 | 24.6 | 62.0 | 52.6 |
| Income burden (Percent of household income earned, weighted by number of people in household) | | | | |
| Low income burden | 34.2 | 33.3 | 3.3 | 5.3 |
| High income burden | 11.0 | 8.7 | 51.2 | 57.9 |

| | | | | |
|---|---|---|---|---|
| Time pressures | | | | |
| Number of hours of constrained time each week (based on job, commuting, household chores, child care, volunteer work, clubs, and church hours) | | | | |
| Less than 40 hrs. | 7.9 | 5.2 | 5.1 | 5.0 |
| 70 or more hours | 23.0 | 20.8 | 31.5 | 40.0 |
| Always feel rushed | 12.7 | 15.6 | 10.2 | 5.0 |
| Never have time-on-hands | 32.8 | 29.9 | 22.8 | 10.0 |
| Involved in too many things | 12.3 | 13.2 | 17.8 | 20.0 |
| Worn out at end of day | | | | |
| Every day | 6.9 | 6.5 | 5.9 | 0.0 |
| Often | 22.1 | 23.4 | 15.0 | 15.0 |
| Role satisfaction | | | | |
| Feeling about job | | | | |
| Like job | 72.9 | 64.9 | 67.7 | 40.0 |
| Unqualified like | (67.0) | (59.7) | (58.9) | (40.0) |
| Qualified like | (5.9) | (5.2) | (8.8) | (0.0) |
| Just has to be done; like some things about it, dislike others | 20.2 | 26.0 | 25.3 | 45.0 |
| Dislike job[a] | 6.9 | 9.1 | 7.1 | 15.0 |
| Feeling about housework | | | | |
| Like housework | 53.7 | 50.0 | --[b] | -- |
| Unqualified like | (40.4) | (32.9) | | |
| Qualified like | (13.3) | (17.1) | | |
| Just has to be done; like some things about it, dislike others | 39.9 | 42.1 | | |
| Dislike housework[a] | 6.4 | 7.8 | | |
| Feeling about main role (For men, their jobs. For women, their jobs or housework, depending on which they consider more important to them) | | | | |
| Like main role | 71.9 | 66.2 | 67.7 | 40.0 |
| Unqualified like | (61.6) | (53.2) | (58.9) | (40.0) |
| Qualified like | (10.3) | (13.0) | (8.8) | (0.0) |
| Just has to be done, etc. | 21.7 | 24.7 | 25.3 | 45.0 |
| Dislike main role[a] | 6.4 | 9.1 | 7.1 | 15.0 |
| Importance of job versus housework[c] | | | | |
| Job more important | 60.6 | 61.0 | -- | -- |
| Equally important | 11.3 | 14.3 | | |
| Housework more important | 28.1 | 24.7 | | |
| Life satisfaction[d] | | | | |
| Very high | 15.7 | 11.7 | 12.3 | 10.0 |
| High | 17.2 | 14.3 | 21.3 | 15.0 |
| Moderate | 37.3 | 37.7 | 49.0 | 70.0 |
| Low | 21.1 | 22.1 | 13.8 | 5.0 |
| Very low | 8.8 | 14.3 | 3.6 | 0.0 |

[a] Unqualified and qualified dislike are combined here.
[b] Question not asked to men respondents.
[c] "Apart from the money, which do you think is more important to you personally--your housework or the work for which you are paid?"
[d] "On this scale from 1 to 10, where 1 means the worst life you could expect and 10 means the best life you could expect, which number would you give your life in the past year?" Here, very high = 10, and very low = 1 - 4.

We decided to look more closely at married clerical women with children, a group with presumably heavy role burdens. Are they more pressured than clerical women who have fewer family responsibilities, or than married mothers in other occupationsμ (1) Compared with other clerical women, married clerical mothers are more family oriented. Many have part-time jobs rather than full-time ones. Adding in their family activities, they are actually busier than other clerical women, having more objective time constraints and also feeling worn out more often. But they are notably happier with their jobs and lives. This is a profile of busy women who have made accommodations so that their employment and family responsibilities mesh. (2) Compared with married mothers with other jobs, clerical mothers have fewer pressures; they work less, have fewer children, and have older children. They often view their family roles as more important than their jobs. Nevertheless, they remain less happy with their roles and lives than other married mothers.

In sum, multiple roles do increase time constraints for clerical women, but married clerical mothers keep their job commitments modest and end up being *happier* than other clerical women. Multiple roles are less demanding for clerical mothers than other working mothers, but the clericals are nevertheless *less happy* with their roles and lives. This is a common theme for clerical workers in Detroit; clerical women in other family statuses (married without children; nonmarried) are also less happy than comparable women in other jobs. What is it about Detroit clerical women's jobs, lives, or psyches that makes them less contentμ

## Clerical men

*Clerical men have fewer job and family responsibilities than most employed men, but they are very unhappy with their jobs.*

Compared with all employed men, Detroit clerical men have more regular schedules, fewer work hours, and fewer family ties and responsibilities (Table 4).[5] Their lower family involvement is partly due to age; 71Δ are ages 18 to 34, compared with 41Δ of all employed men. Clerical men have more commitments during their leisure hours than other men, and overall they end up with as many time constraints. The most striking feature of clerical men is how unhappy they are with their jobs; they rank decidedly lowest of all occupational groups for job satisfaction. Despite this, they are quite satisfied with life in general; possibly their leisure activities help to offset job dissatisfaction.

## Summary

This review shows that clerical women and men in Detroit do not have heavy role burdens; in fact, they have fewer responsibilities and commitments than employed people generally. But they tend to be more dissatisfied with their job and family situations than other workers.

Dissatisfaction with roles is associated with poor health (ref. 24). Dissatisfaction may be a risk factor, increasing a person's vulnerability to illness and injury, or exacerbating current conditions. Social selection is also relevant, if poor health forces people to change their roles and they are very unhappy with the new ones. Exactly how dissatisfaction is routed to poor health outcomes, or vice versa, is not known.

Does the unhappiness of Detroit clerical women partly explain their poor health profile? Is the unhappiness of Detroit clerical men the result of poor health, rather than the cause? Although we have just attached a social causation hypothesis to women and a social selection hypothesis to men, both processes can operate for each sex.

The Detroit data show a conjunction of poor health and role dissatisfaction among clerical workers. This is a provocative result, but a scientifically incomplete one. The study cannot reveal just why clerical workers are unhappy, or exactly how dissatisfaction jeopardizes health. The answers to these questions must be elucidated by other studies.

## COMPARING NATIONAL, DETROIT, AND FRAMINGHAM RESULTS

The Framingham Heart Study found high rates of coronary heart disease among clerical women, compared with other occupation groups (ref. 5). Two subgroups of clerical women were responsible for the high rates: those who had borne children (mothers), and those with husbands in blue-collar jobs. Clerical women with both characteristics had especially high CHD rates. Looking at all clerical women, the ones who developed heart disease tended to have non-supportive employers, low job mobility (few job changes in the past 10 years), a tendency to suppress hostility rather than express it, and few personal worries.[6] Looking at clerical mothers, the same psychosocial risk factors appeared. (Comparable analysis for clericals married to blue-collar husbands is not reported. Nor are analyses of risk factors within the other two occupational groups of women.) These results suggest that Framingham clerical women experienced abnormally high stresses from a combination of factors--an unfulfilling work situation plus childrearing duties plus financial pressures.[7] Moreoever, their tension remained bottled up. In conceptual terms, these risk factors refer to role burdens, role satisfactions, and psychological traits. It is the combination of factors that boosted their risks; taken alone, each factor is much less powerful.[8] In contrast to the women, clerical men in Framingham had very low incidence of CHD, lower than other occupational groups of men. No additional analysis of risk factors for clerical men was done.

A brief summary of our NHIS and Detroit results: American clerical women have a very good health profile--in fact, the best one overall compared with other occupations. American clerical men have a relatively poor profile, especially compared with other white-collar occupations. The Detroit study concurs with NHIS for clerical men, but it shows a rather poor profile for clerical women. Positing that role burdens and satisfactions may be important risk factors, we have looked at aspects of Detroit clericals' jobs and family ties. Their role burdens are actually less than employed people in general. But they are clearly less satisfied with their roles.

Clerical women in Framingham and the United States

Despite many compatible results for the three surveys, one difference is very troubling--the different health profiles of Framingham and American clerical women.[9] To understand the contradiction, we need to recognize how the Framingham Heart Study and the National Health Interview Survey differ. First, the health indicators are very different. Framingham concentrates on a specific health problem (coronary heart disease), whereas NHIS has more general measures of morbidity. Also, Framingham studies incidence of disease over an 8-year prospective period, while NHIS asks about current health status and health events in the past year. Second, Framingham is a community study, while NHIS is a national one. Third, the Framingham sample is middle-aged adults (45 to 64 in 1965-67), while NHIS includes adults of all ages. Fourth, there are large differences in how employment characteristics are defined. (1) For Framingham, "clerks and kindred workers" includes sales jobs. Most reports abbreviate this to "clericals," which further masks the presence of sales workers. For NHIS, the Census category "clerical and kindred workers" does not include sales jobs. ("Sales workers" are a separate Census category, and we do not study them in this paper.) In addition, the occupation categories used for Framingham emphasize social status (see ref. 5), whereas the Census Bureau categories used for NHIS emphasize objective job characteristics. (2) Definitions of employment status also differ: Framingham women are coded as "working" if they were employed outside the home for more than half their adult life up to 1967. (See ref. 5 for further restrictions on the definition.) Their employment status during the study period (1965-75) is not considered; many were certainly retired by then. "Working" therefore, means having substantial employment experience before the study period started. In contrast, NHIS results refer to currently employed people; the clerical women are all employed at the time of interview, and their prior employment experience is not relevant.

How can these four factors influence the results? We suspect that the first three are not very important, but that the last one is critical. First, heart disease is a common chronic problem, and it usually limits activities and requires medical care. Groups with high heart disease incidence and prevalence are likely to have a poor health profile in NHIS, even if the survey lacks specific heart disease indicators. Second, Framingham was selected as a typical community, so it is not likely that the clerical women there experienced special risks that most American clerical women did not. Third, the wider age range for NHIS could mask a poor health profile for middle-aged clericals. To assess this, we looked at NHIS data for clerical women ages 45 to 64 and found that they match the overall profile, having very low rates of chronic limitation, restricted activity, and work loss.[10]

Fourth, we believe that the different results signal a cohort effect. It arises from the different employment characteristics of Framingham and NHIS clerical women: the Framingham women are ages 45 to 75 during the study period (1965-75). They worked mainly during the 1940's-1960's, and they worked a large fraction of those years. Female employment was not encouraged in those decades. Women who worked often did so because of financial pressures rather than choice, and they found limited social supports at their jobs or at home. The Framingham results point to women who were in dead-end clerical jobs for many years and who also had children at home, nonsupportive bosses, financial pressures, and few outlets for stress. The Framingham situation was probably typical for American clerical women in the 1940's-1960's. By contrast, in the past 2 decades female employment has become more common, more encouraged, and more rewarding. Women with a wide variety of family attachments, reasons for work, and coping skills have entered the labor force. Contemporary clerical women probably have more pleasant and supportive situations than were typical in earlier decades. If so, this reduces their risks of stress-related disease, and it is reflected in a good contemporary health profile.

What does this imply for research on women's health? What scientists need to look for is women with combinations of stresses--from unfulfilling jobs, child care responsibilities, financial pressures, and few vents for their emotions. They may be clericals, professionals, service workers, or any other occupation. The Framingham study's contribution is to identify a certain combination of roles and psychological makeup that leads to coronary heart disease. Those risks happened to be very prevalent among women with clerical (including sales) jobs in the 1940's to mid-1960's.

Clerical men

There are contradictory profiles here too. The Framingham men have a good profile, while American men do not. We suspect this is due to two factors--the occupation coding and the health indicators studied. First, when clerical and sales men are combined in Census data, about 50% of them are clericals and 50% are sales; we assume this split is roughly true for Framingham. CHD risks may be quite different for the two groups. Even more important is how strongly prior health affects men's choice of clerical and sales jobs. The Framingham results may refer to men with very diverse health risks and selection forces, while NHIS results refer to a more homogeneous group of (just clerical) men. Second, the Framingham study selected people free of CHD in 1965-67 and then assessed CHD incidence over the next 8 years. If the focus had been CHD prevalence and all clerical men were surveyed, we might discover a poor profile--more compatible with that in NHIS. The fact that the Framingham men chosen do not develop much heart disease actually buttresses our notions of social selection--that the reason for the poor health profile of American (and Detroit) clerical men is prior health problems that made them take such jobs, more than risks they encounter while they are clerical workers. In sum, we suspect that the good profile of Framingham men is due to the inclusion of sales workers and the restriction to initially healthy people.[11]

## CONCLUSION

The elevated incidence rates of coronary heart disease (CHD) found among Framingham clerical women has spurred intense public interest in clerical workers' health. In this paper, we have found a different situation for American clerical women; they show a very good health profile. We have also looked at American clerical men; they have a very poor health profile compared with other white-collar workers. Profiles of clericals in Detroit were also studied.

Such profiles are interesting and they offer solid descriptive information. But they only hint at risk factors--the aspects of job, family, lifestyle, psyche, and physical environment that lead to poor health outcomes. The Framingham study does identify some risk factors among clerical women. These refer to job (nonsupportive boss, few job changes), family (children, blue-collar husband), and psyche (inability to express hostility, few personal worries). For the Detroit survey, we looked at role burdens and satisfactions among clerical women and clerical men, and we discovered that they are more dissatisfied with their job and family roles than other occupational groups. Dissatisfaction is probably a risk factor for the poor health we find among Detroit clericals.

Typically, risk factor research takes a large array of items and winnows them, extracting the ones that are significantly related to poor health when the others are controlled. The result is a small list of factors, each one having a link with poor health. This approach yields important information, but it may miss some powerful effects. It is very likely that *combinations* of factors have unusually strong impacts on health, far greater than we would predict by just adding up their separate effects. In other words, certain mixes of job hazards, role burdens, and psychological traits boost risks of chronic disease greatly. These show up statistically as interaction effects. This is, in fact, the striking feature of Framingham's result: for clerical women there, the mix of job plus motherhood plus blue-collar husband made for very high risks of heart disease.[12] Further analysis suggested that these women had unsatisfactory jobs, and that they tended to suppress anger. A picture of pressured women emerges--women who had unfulfilling jobs for many years, childrearing duties, and also financial pressures. Future research should look for such combinations that pose unusually high risks of disease or injury. One place to look, especially for women, is at role burdens, role dissatisfactions, and weak coping skills. (For more discussion of women, work, and stress, see ref. 26.)

The ideal data for risk factor research are prospective, with posited risk factors measured at the outset and morbidity events thereafter. By contrast, when the predictors and health are measured at the same time, results are very difficult to interpret. An association between a predictor (X) and health (Y) might reflect social causation (X→Y) or social selection (Y→X). Only in the first instance is X a risk factor. Our analysis of clerical men offers clues that social selection operates strongly for them--that poor health forces some men to take clerical jobs, and they become unhappy about that role. Until recently, health scientists seldom discussed social selection, and they rarely tried to measure it. This has changed, and researchers are now keenly interested in measuring the relative importance of causation and selection. This is especially critical for occupational health research, since health certainly affects whether people become employed and what kind of jobs they have.

There are some sound reasons for studying the health of clerical women. They are a large occupational group, and their job situations are changing rapidly. But researchers should not expect to find them in generally poor health; in fact, American clerical women in the 1970's have a very good health profile. We suspect the Framingham results point to pressures that were especially common for clerical women during the 1940's-mid 1960's, and that those risks have lessened for contemporary clerical workers and are also less concentrated in that occupational group now. The Detroit survey does show that clerical women there have a poor health profile and that they are less satisfied with their roles than employed women generally. We think these clerical women are not typical of American clericals. The truly important aspect of the Detroit results is the link between role dissatisfaction and poor health. Regardless of a person's occupation, prolonged unhappiness with roles may jeopardize health.

Many changes in American society have made it easier for women to work, especially to have full-time uninterrupted employment. But inflation and discrimination continue to blunt women's pleasures and their opportunities for success. Scientists may not be able to determine if life pressures have diminished for contemporary working women compared with their peers several decades ago, or if pressures are less focused on clericals now than before. Instead, the goal should be to identify aspects of roles and psyche that most damage women's health. These links between risk factors and poor health outcomes are enduring ones. All occupational groups must be considered; whether the risks are particularly high for contemporary clerical women should be a research question rather than a presumption.

ACKNOWLEDGMENT

Funds for collection and analysis of the Health in Detroit data came from the Center for Epidemiologic Studies, National Institute of Mental Health (Research Grant MH29478).

FOOTNOTES

1. Published data from NHIS are seldom adequate to show this, because tables with rates by job class do not also show rates for nonemployed groups. (For 1961-63, rates are not given for people outside the labor force. For 1975-76, rates are given for unemployed people, but not for people outside the labor force.) Our summary is based on three sources: (1) unpublished 1972-73 NHIS tables provided to the author, which have rates for not-in-labor force persons, (2) published 1975-76 NHIS tables, which have rates for unemployed people, and (3) Health in Detroit rates for unemployed and not-in-labor force groups.

2. These low rates appear for 1975-76 and 1961-63; rates are average for 1972-73.

3. One result disagrees at first glance. In the 1961-63 NHIS, clerical men have average rates of limiting heart conditions. Farmers/farm managers have the highest rates, but their older age distribution is certainly a factor.

4. This is strongly true for 1972-73, but it does not appear in the 1961-63 data.

5. Compared with clerical women, the clerical men have more variable schedules and more work hours.

6. Having few personal worries might also reflect retention of emotions, rather than a troublefree life.

7. Several interpretations of the "blue-collar husband" effect are possible: (1) financial pressures forced wives to work when they preferred not to, or forced them to have full-time uninterrupted employment when they preferred part-time occasional employment, (2) blue-collar husbands offered little help with child care and housework. (See ref. 26 for additional discussion.)

8. For the whole sample of Framingham women, univariate analyses show that these items are not significantly related to CHD: blue-collar husband, personal worries, nonsupportive boss. Two items have a significant link in

univariate analysis but not in multivariate analysis; few job changes (significant for ages 45 to 54 only), having children. Only one item is significant in multivariate analysis: suppressed hostility. (See ref. 5, 27.)

9. The difference between Detroit and American clerical women is also perplexing. The Detroit survey has a similar design and many similar health items to NHIS. As stated earlier, the most plausible reason is that clerical women in Detroit are not representative of American clericals.

10. This is based on 1972-73 tabulations; the 1975-76 and 1961-63 data are not available by age.

11. These two factors are less critical for women. First, when U. S. clerical and sales women are grouped together, about 84% are clerical and only 16% are sales. Second, clerical jobs are probably less selective for healthy women; if so, the CHD-free clericals chosen for the study are quite representative of all clericals in Framingham.

12. Unfortunately, published results for Framingham do not let us show this statistically. To do so, we need the three factors' additive effects, then the extra interaction effect for women who have all three characteristics.

REFERENCES


1 V. E. Giuliano, The mechanization of office work. Scientific American, (September, 1982) 148-164.

2 U. S. Department of Labor, Bureau of Labor Statistics, Perspectives on working women: A Databook, 1980, Bulletin 2080, Washington, D. C., 1980.

3 J. M. Stellman, Women's Work, Women's Health: Myths and Realities, Pantheon, New York, 1977.

4 Women's Occupational Health Resource Center, WOHRC News, Various issues (School of Public Health, Columbia University, 60 Haven Avenue, B-1, New York, New York 10032) 1982.

5 S. G. Haynes and M. Feinleib, Women, work and coronary heart disease: prospective findings from the Framingham Heart Study. American Journal of Public Health, 70:2 (1980) 133-141.

6 G. A. Gleeson, Selected health characteristics by occupation, vital and health statistics, Series 10, No. 21, National Center for Health Statistics, Washington, D. C., 1965.

7 C. S. Wilder, Selected health characteristics by occupation, United States, 1975-76, Vital and health statistics, Series 10, No. 133, DHHS Publ. No. (PHS) 80-1561, National Center for Health Statistics, Hyattsville, Md., 1980.

8 L. M. Verbrugge, Female illness rates and illness behavior: testing hypotheses about sex differences in health. Women and Health, 4:1 (1979).

9 L. M. Verbrugge, Health diaries. Medical care, 18 (1980) 73-95.

10 L. M. Verbrugge, Health diaries--problems and solutions in study design. Paper presented at the fourth Health Survey Research Methods Conference, Sponsored by the National Center for Health Services Research, National Center for Health Statistics, and Milbank Memorial Fund. Forthcoming in C. Cannell (Ed.), Health Survey Research Methods--Fourth Conference. Research Proceedings Series, National Center for Health Services Research, Hyattsville, Md., 1982.

11 A. C. Marcus and T. E. Seeman, Sex differences in reports of illness and disability: a preliminary test of the fixed role obligations hypothesis. Journal of Health and Social Behavior, 22:2 (1981) 174-182.

12 C. A. Nathanson, Social roles and health status among women: The significance of employment. Social Science and Medicine, 14A (1980) 463-471.

13 D. P. Rice and A. S. Cugliani, Health status of american women, proceedings of the American Statistical Association (Social Statistics Section), 1979, pp. 72-78.

14 L. M. Verbrugge, the Social roles of the sexes and their relative health and mortality. Paper presented at conference on sex differentials in mortality, sponsored by the World Health Organization. Forthcoming in A. Lopez (Ed.), Sex differentials in mortality: trends, determinants, and consequences, World Health Organization, Geneva, 1981.

15 L. M. Verbrugge, Women's social roles and health in P. Berman and E. Ramey (Eds.), Women: a developmental perspective, NIH Publ. No. 82-2298, National Institute of Child Health and Human Development, Bethesda, Md., 1982, pp. 49-78.

16 S. Welch and A. Booth, Employment and health among married women with children. Sex Roles, 3 (1977) 385-397.

17 N. F. Woods and B. S. Hulka, Symptom reports and illness behavior among employed women and homemakers. Journal of Community Health, 5 (1979) 36-45.

18 L. M. Verbrugge, Sex differentials in health. Public Health Reports, (September/October 1982).

19 W. R. Gove, The relationship between sex roles, marital status, and mental illness. Social Forces, 51 (1972), 34-44.

20 R. C. Kessler and J. A. McRae, Jr., The effect of wives' employment on the mental health of married men and women. American Sociological Review, 47 (1982) 216-227.

21 R. L. Repetti and F. Crosby, Women and depression: the protective function of multiple roles. Unpublished manuscript (Dept. of Psychology, Yale University, Box 11A Yale Station, New Haven, CT 06520) 1982.

22 E. Spreitzer, E. E. Snyder and D. L. Larson, Multiple roles and psychological well-being. Sociological Focus, 12 (1979) 141-148.

23 L. M. Verbrugge, Multiple roles and physical health of women and men. Journal of Health and Social Behavior, 1983, in press.

24 L. M. Verbrugge, Work satisfaction and physical health. Journal of Community Health, 7:4 (1982) 262-283.

25 National Institute for Occupational Safety and Health, Potential health hazards of video display terminals, DHHS (NIOSH) Publ. No. 81-129, Cincinnati, 1981.

26 M. A. Haw, Women, work and stress: a review and agenda for the future. Journal of Health and Social Behavior, 23 (1982) 132-144.

27 S. G. Haynes, M. Feinleib and W. B. Kannel, The relationship of psychosocial factors to coronary heart disease in the Framingham Study. American Journal of Epidemiology, 111 (1980) 37-58.

# CLERICAL WORK AND CORONARY HEART DISEASE IN WOMEN: PROSPECTIVE FINDINGS FROM THE FRAMINGHAM STUDY

SUZANNE G. HAYNES
Department of Epidemiology
University of North Carolina
School of Public Health
Chapel Hill, North Carolina 27514

MANNING FEINLEIB
Epidemiology Branch
National Heart, Lung, and Blood Institute
Federal Building, Room 2C-08
Bethesda, Maryland 20205

## ABSTRACT

During the mid-1960's, 350 housewives, 387 working women (women who had been employed outside the home over one-half of their adult years), and 580 men participating in the Framingham Heart Study were administered an extensive life history questionnaire. Approximately 35% of Framingham working women had been employed in clerical occupations during their working lives. The respondents were 45 to 64 years of age and were followed for the development of coronary heart disease (CHD) over the ensuing 10 years.

Working women did not have significantly higher incidence rates of CHD than housewives (8.5 versus 7.1%, respectively). However CHD rates were almost twice as great among women holding clerical jobs (12.0%) as compared to housewives. Compared to nonclerical workers, clerical women developing CHD were more likely to have had nonsupportive bosses, limited job mobility, and suppressed hostility. In addition, the risks of CHD were greater among working women with children. Clerical workers who had children and were married to blue-collar workers were three times more likely to develop CHD than similar nonclerical workers (21.3 versus 6.0%, respectively). These findings suggest that family pressures along with an undesirable work environment may lead to an increased risk of CHD among clerical working women.

## INTRODUCTION

The growing participation of women in the workplace has brought fears that women will lose their survival advantage over men and will have increasingly higher mortality rates from chronic diseases such as coronary heart disease (CHD). Contributing to these fears is an unsubstantiated assumption that men live fewer years than women because they work outside the home.

During the past 30 years, the number of women in the United States labor force has risen sharply. In this period, the proportion of women in the labor force has increased from 28% in 1950 to 51% (ref. 1, 2). Most of this growth has resulted from an influx of married women into the labor force (ref. 3).

In order to examine the effect of employment on the cardiovascular health of women, the present study followed working women, housewives, and men participating in the Framingham Heart Study over a 10-year period for the development of CHD. Working women were classified into white-collar, blue-collar, and clerical occupations, the latter of which comprised over 35% of the Framingham working population. While results from the 8-year follow-up study have been published previously (ref. 4), the current study provides additional follow-up (2 years) on morbidity and mortality from CHD in the Framingham cohort.

## METHODS

Between 1965 and 1967, an extensive psychosocial questionnaire was administered to a sample of men and women in the Framingham cohort undertaking their eighth or ninth biennial medical examinations. The present analysis includes the 350 housewives, 387 working women, and 580 men, aged 45 to 64 years, who were free of CHD at the time of the examinations. Although persons 65 years of age and over were also included in the original study, the present analysis was restricted to individuals in their employment years. A comprehensive description of the characteristics of this sample of the Framingham cohort has been reported previously (ref. 5). In most respects, the sample under study appears representative of the entire study population.

Women who indicated they had been employed outside the home for over one-half of their adult years (age 18+) were designated "working women"; otherwise they were classified as "housewives." Thus, a working woman 50 years of age would have worked the full-time equivalent of at least 15 years outside the home.

Occupation, as defined by one's usual lifetime work, was grouped into the following six categories according to the Warner index of status characteristics (ref. 6): professionals, proprietors and managers, businesspeople, clerks and kindred workers, manual workers, and protective and service workers. The first three groups were designated white-collar occupations, the last two groups were blue-collar occupations, and clerical jobs were considered separately.

Twenty psychosocial scales were examined in this study. A complete description of their content, including reliability coefficients and interscale correlations, may be found in a previous publication (ref. 5). The scales were grouped in five categories: behavior types, situational stress, anger reactions, somatic strains, and sociocultural mobility. A family responsibility scale was developed to account for marital status and the number of children in the family. Respondents were scored as single (1); evermarried, no children (2); ever-married, 1 to 3 children (3); or ever-married, 3+ children (4).

The entire study group was followed for the development of CHD over a 10-year period. CHD was diagnosed if, upon review of all clinical and examination data, a panel of investigators agreed that a myocardial infarction, coronary insufficiency syndrome, angina pectoris, or CHD death had occurred. Definitions of these clinical manifestations of CHD have been presented elsewhere (ref. 7). The systolic and diastolic pressures, serum cholesterol (mg/100 ml), cigarette smoking, and relative weight measures used in this study have been described previously (ref. 5).

Statistical differences in coronary incidence rates were determined by a two-sided Chi-square test or the $Z$-statistic for testing differences between two proportions (ref. 8). The direct method of age adjustment, using all Framingham men and women (ages 45 to 54 and 55 to 64 years) in this study as the standard population, was used to test whether observed differences in CHD rates were due to differences in the age distributions between groups. The associations were unaffected by the adjustment for age. Thus, unless otherwise stated, unadjusted incidence rates for the entire age group 45 to 64 years will be presented throughout the analysis.

## RESULTS

### Employment patterns

Over one-third of all working women had been employed in clerical and kindred occupations during their working years. Secretaries, stenographers, bookkeepers, bank clerks and cashiers, and sales personnel made up the majority of these positions. Although equal proportions of working women and men were employed in white-collar jobs (20%), more women (37%) were employed in clerical occupations than men (18%), and fewer women (43%) were employed in blue-collar jobs than men (62%) ($p < .001$, comparing occupations of men and women). The majority of men with white-collar occupations were graduate-degree professionals (lawyers, doctors, dentists, etc.) or business managers, while most women professionals were teachers, nurses, or librarians.

### Incidence rates of CHD

Fig. 1 presents incidence rates of CHD over the 10-year period among housewives, working women, and men aged 45 to 64 years. Data were also analyzed separately for working women, as previously defined, who were currently employed at the time of the study. All working women were included in the ever-employed group.

Employment status, *per se*, did not significantly affect the risk of developing CHD in women. Incidence rates were only slightly higher among the ever-employed working women than among housewives (8.5 versus 7.1%, respectively). The incidence rate of CHD among these working women was lower than the rate for men, which was about 15% ($p = .003$).

Fig. 2 shows incidence rates of CHD among working women and men according to the usual occupation held during the working years. Among women, clerical workers were almost twice as likely to develop coronary disease as either white- or blue-collar workers. The incidence rate of CHD among women clerical workers (12.0%) was higher than the rate among housewives (7.1%, $\underline{p}$ = .075).

Among men, an entirely different pattern was observed, with higher rates occurring among white-collar workers (21.6%) and lower rates occurring among clerical (11.5%) and blue-collar (14.4%) employees ($\underline{p}$ = .097). Only among clerical workers were the rates of coronary disease greater in women than in men, although this difference did not achieve statistical significance.

Age-adjusted coronary rates were examined among working women and housewives according to marital status. No significant differences were observed among housewives who were married and housewives who were widowed, divorced, or separated (WDS) (6.2 versus 10.3%, respectively). Married and WDS working women had similar age-adjusted rates of CHD (8.9 and 9.4%, respectively), while single working women exhibited the lowest rate of coronary disease (5.7%).

Since women who had ever married were at greater risk of developing CHD than single women, the effect of having children on CHD was also examined. Among working women, the incidence of CHD rose as the number of children increased (Fig. 3). Working women with three or more children (11.0%) were more likely to develop CHD than working women with no children (7.8%) or than housewives with three or more children (5.7%), although these differences did not reach statistical significance ($\underline{p}$ = .23). CHD rates were similar among housewives with 1 to 2 or 3+ children.

Although one would expect working women to be equally affected by family responsibilities, the relationship of these responsibilities to CHD incidence was examined among clerical and nonclerical working women (Fig. 4). Surprisingly, single or married clerical workers without children were at no greater risk of developing CHD than other workers. However, clerical workers who had ever married and had children were over twice as likely to develop CHD than nonclerical workers in the same situation (15.4 and 6.9%, respectively, $\underline{p}$ = .057). Thus, the increased risk of CHD observed among women employed in clerical jobs occurred only among women with children.

Economic pressures resulting from increased family size could have motivated women to seek employment outside the home. Pressures associated with a low socioeconomic status might then explain the higher incidence rate of CHD among working women with children. Although measures of family income were not available, the occupation of a woman's past or present husband was examined. For these comparisons, men employed in white-collar and clerical occupations were combined.

The risk of developing CHD did increase among clerical mothers married to blue-collar workers (Fig. 5). Among working women with blue-collar husbands, clerical mothers were over three times more likely to develop CHD than nonclerical mothers (21.3 and 6.0%, respectively, $p = .004$). Among working mothers married to white-collar workers, clerical work posed no increased risk of CHD. Likewise, clerical working wives with no children were at similar risk of developing CHD as comparable nonclerical wives, whether married to a white-collar (16.7 versus 12.5%, respectively) or to a blue-collar husband (4.2 versus 5.0%, respectively). Thus, the increased risk of CHD among clerical working women lies solely among those workers who were married, who had raised children, and who were wives of blue-collar workers.

## Differences in standard coronary risk factors

Fig. 6 presents mean levels of the standard coronary risk factors measured between 1965 and 1967 among the various occupational groups. The risk factors included age, systolic and diastolic blood pressure, serum cholesterol, and cigarette smoking. In addition, glucose intolerance, relative weight, and the proportion of persons on antihypertensive medication were also compared.

Mean levels of the five standard coronary risk factors were similar among housewives and working women, regardless of occupation. Blue-collar working women were more obese than clerical workers (105.4 versus 101.7, respectively, $p < .05$), who had the lowest mean relative weight of all the women examined. Prevalence rates of hypertension (SPB $\geq$ 160 or DBP $\geq$ 95) among women did not vary by employment or occupational status.

Men, on the other hand, had significantly higher levels of cigarette consumption and lower levels of serum cholesterol than working women or housewives. Mean levels of diastolic blood pressure were also significantly higher among men than women.

## Risk factors for CHD among clerical workers

In a previous report from Framingham (ref. 7), several psychosocial scales were associated with the development of CHD in women, depending upon employment status. Since clerical workers were at greater risk of developing CHD than other workers or housewives, the psychosocial as well as standard coronary risk factors were examined in this group of women.

In a univariate analysis, clerical workers who had developed CHD were more likely to suppress hostility (in terms of the anger-in, anger-out, anger-discuss scales), to have nonsupportive bosses, to report fewer personal worries, and to experience fewer job changes over a previous 10-year period than clerical workers remaining free of CHD ($p \leq .05$). In addition, we have shown that family responsibilities (number of children), combined with a husband's occupation, were associated with CHD among the clerical working women.

In order to determine the independent effect of these variables over and above the standard coronary risk factors, each was included in a multivariate logistic regression analysis (ref. 9). As seen in Table 1, the anger-discuss (negative), nonsupport from boss, and job changes (negative) scales remained independent predictors of CHD in the multivariate analysis, after controlling for the other risk factors (Analysis I). Husband's occupation, *per se*, was not independently related to the development of CHD in these women (Analysis II). When both number of children and husband's occupation were included in the regression model (Analysis III), the statistical significance of the number of children was reduced.

Upon examination of the correlation between husband's occupation and the number of children, a significant correlation between husband's occupation and number of children was observed, particularly among the cases ($r=.571$ in cases; .021 in non-cases). This suggested that there may be some interaction between family size and a husband's occupation in these risk models. In order to explore this possibility further, an additional regression model was run in which number of children, husband's occupation (1=white collar, 2=clerical, 3=blue collar) and the interaction of the two (number of children x husband's occupation) were included with the other variables. In this analysis, the interaction term was a significant risk factor for CHD ($p < .05$, $Z = 2.00$), suggesting that the risk of developing CHD among clerical workers rose with an increasing number of children, but only in families headed by lower occupation status men. In sum, family responsibilities, in addition to an undesirable work environment (remaining in a job with a nonsupportive boss while not discussing one's anger), significantly increased the risk of coronary heart disease among the clerical workers, independently of the standard coronary risk factors.

## DISCUSSION

Of the occupations examined in this study, clerical work was associated with the greatest risk of CHD among women. Since over one-third of the female workers in the U. S. are employed in clerical jobs (ref. 10), reasons for this excess risk require further examination. Unfortunately, few epidemiologic data are available in the U. S. on cardiovascular morbidity or mortality rates among women according to occupation. A 1977 NIOSH study of 130 occupations found the incidence of stress-related diseases to be high among secretaries, with clerical rates ranking second highest among all the occupations examined (ref. 11).

Results presented by Verbrugge also showed that clerical women in the Health in Detroit study reported more chronic health problems, more daily symptoms, more job limitations, and more prescription drug use than all employed women (ref. 12).

The health effects of clerical work are more often observed from morbidity rather than mortality rates, since few of the clerical workers in Framingham died of cardiovascular disease. Furthermore, mortality statistics from Scandinavia and England indicate that clerical workers, as a whole, die from CHD at expected rates, based on national averages (ref. 13). In addition, as noted in the Verbrugge study, health problems among clerical workers appear more often in local studies rather than national studies, such as the Health Interview Study (ref. 12). These differences may be due to geographic differences, cohort effects, or to the healthy worker effect (ref. 14), which is more likely to be observed in national studies.

The association between clerical work and CHD incidence in women could be explained, in part, by the distribution of standard coronary risk factors by employment and occupational status. However, mean levels of blood pressure, serum cholesterol, cigarette smoking, and glucose intolerance in Framingham were similar among housewives and working women, regardless of occupation. These findings are consistent with other national and population-based surveys (ref. 15, 16, 17, 18, 19).

In Framingham, suppression of hostility coupled with a nonsupportive boss and a lack of job mobility were associated with the incidence of coronary heart disease among clerical working women. Many of these behaviors appear to be related to employment; that is, they are the result of working outside the home or the self-selection of certain persons into the work force. Studies by Harburg and others among employed persons in Detroit showed that white women were more likely than white men to suppress hostility (more anger-in and less anger-out) when confronted with an arbitrary boss (ref. 20). These findings are consistent with observations that women clerical workers may experience several forms of occupational stress, including a lack of autonomy and control over the work environment, underutilization of skills, and lack of recognition of accomplishments (ref. 10).

The excess risk of CHD observed among women employed in clerical jobs occurred only among women with children and among women married to blue-collar workers, suggesting that economic pressures may also have affected the decision or necessity to work. Since the risks of CHD did not increase among non-clerical working mothers with blue-collar husbands, the exact meaning of these results is unclear. The occupational status of one's spouse reflects not only an economic status, but also certain lifestyle behaviors and attitudes not measured in this study.

In conclusion, although employment, per se, was not associated with the incidence of coronary heart disease in women, behaviors and situations related to employment were associated with CHD among some working women. Working women who had ever married, had raised children, and had been employed in clerical work were at increased risk of developing CHD. Job-related characteristics associated with CHD among clerical women included suppressed hostility, a non-supportive boss, few job changes over a 10-year period, and family responsibilities. These situations may be the product of one or more of the following factors: the particular working environment for clerical occupations, self-selection of certain personalities into the labor force, or economic stress. Whatever the origins of these situations, the findings suggest that women in clerical occupations, coupled with family responsibilities, may be at higher risk of developing CHD.

REFERENCES

1 U. S. Dept. of Labor, Women's Bureau, Changes in Women's Occupations 1940-1950, U. S. Government Printing Office, Women's Bureau Bulletin 253, Washington, DC, 1954.
2 U. S. Dept. of Labor, Bureau of Labor Statistics, Employment and Earnings, U. S. Government Printing Office, Washington, DC, 1981.
3 Lipman-Blumen, Demographic trends and issues in women's health, in V. Olesen (Ed.), Women and Their Health: Research Implications for a New Era, U. S. Government Printing Office, DHEW Publication No. (HRA) 77-3138, Washington, DC, 1977.
4 S. G. Haynes and M. Feinleib, Women, work, and coronary heart disease: Prospective findings from the Framingham Heart Study, Am. J. Public Health, 70 (1980) 133-141.
5 S. G. Haynes, S. Levine, N. Scotch and others, The relationship of psychosocial factors to coronary heart disease in the Framingham Study I. Methods and risk factors. Am. J. Epidemiol, 107 (1978) 362-383.
6 W. L. Warner, M. Meeker and K. Eells, Social Class in America, Science Research, New York, 1949.
7 S. G. Haynes, M. Feinleib and W. B. Kannel, The relationship of psychosocial factors to coronary heart disease in the Framingham Study III. Eight-year incidence of coronary heart disease. Am. J. Epidemiol, 111 (1980) 37-58.
8 J. L. Fleiss, Statistical Methods for Rates and Proportions, John Wiley and Sons, New York, 1973.
9 S. H. Walker and D. B. Duncan, Estimation of the probability of an event as a function of several independent variables. Biometrika, 54 (1967) 167-179.
10 J. M. Stellman, Occupational health hazards of women: An overview, Prev. Med. 7 (1978) 281-293.
11 M. J. Smith, A review of NIOSH psychological stress research, in Proceedings of a Conference on Occupational Stress, U. S. Government Printing Office, DHEW Publication No. (NIOSH) 78-156, Cincinnati, 1978.
12 L. M. Verbrugge, Women's Social Roles and Health, in P. Berman and E. Ramey (Eds.). Women: A Developmental Perspective, NIH Publ. No. 82-2298, National Institute of Child Health and Human Development, 1982, pp 44-78.
13 Editorial, Women, work, and coronary heart disease, Lancet II (1980) 76-77.

14 A. J. McMichael, S. G. Haynes and H. A. Tyroler, Observations on the evaluation of occupational mortality data. J. Occup. Med, 17 (1975) 128-131.
15 U. S. Department of Health, Education, and Welfare, Public Health Service, National Center for Health Statistics, Hypertension and Hypertensive Heart Disease in Adults--United States, 1960-1962. Vital and Health Statistics, Series 11, No. 13, U. S. Government Printing Office, Washington, DC, 1966.
16 L. S. Hauvenstein, S. V. Kasl and E. Harburg, Work status, work satisfactions, and blood pressure among married black and white women. Psychol. Women Q. 1 (1977) 334-349.
17 M. H. Mushinski and S. D. Stellman, Impact of new smoking trends on women's occupational health. Prev. Med, 7 (1978) 349-365.
18 T. D. Sterling and J. J. Weinkam, Smoking characteristics by type of employment, J. Occup. Med., 18 (1976) 743-754.
19 J. Slack, N. Noble, T. W. Meade, and W.R.S. North, Lipid and lipoprotein concentrations in 1,604 men and women in working populations in northwest London, Br. Med. J, 2 (1977) 353-356.
20 E. Harburg, E. H. Blakelock and P. J. Roeper, Resentful and reflective coping with arbitrary authority and blood pressure, Psychosom. med, 41 (1979) 189-202.

TABLE 1
Multiple logistic regression of the 10-year incidence of CHD among married clerical working women Aged 45-64 years.

| variable | ANALYSIS I | | ANALYSIS II | | ANALYSIS III | |
|---|---|---|---|---|---|---|
| | Standardized coefficient | Z | Standardized coefficient | Z | Standardized coefficient | Z |
| Age | .069 | .20 | .061 | .16 | .160 | .40 |
| Diastolic blood pressure (mmHg) | .461 | 1.40 | .758 | 1.91[a] | .833 | 1.89[a] |
| Number of cigarettes smoked/day | .054 | .17 | .077 | .22 | .219 | .61 |
| Serum cholesterol (mg/100mL) | .330 | 1.05 | .491 | 1.35 | .507 | 1.32 |
| Anger-discuss | -1.220 | -2.26[b] | -.173 | -2.50[b] | -1.878 | -2.59[b] |
| Nonsupport from boss | .581 | 2.43[b] | .684 | 2.44[b] | .784 | 2.57[b] |
| Job changes in past 10 years | -1.514 | -1.88[a] | -.166 | -1.91[a] | -1.660 | -1.84[a] |
| Number of children (0,1 = 1,2, 2 = 3+) | .675 | 2.19[a] | | | .636 | 1.74[a] |
| Husband's occupation (1 = white-collar, 2 = clerical, 3 = blue-collar) | | | -.056 | -.14 | -.126 | -.30 |

[a] $.05 < p \leq .10$.

[b] $p < .05$.

FIGURE TITLES

Fig. 1. Ten-year incidence of CHD among housewives, working women, and men aged 45 to 64 years.

Fig. 2. Ten-year incidence of CHD among working women and men aged 45 to 64 years by occupational status.

Fig. 3. Ten-year incidence of CHD among housewives and working women aged 45 to 64 years by number of children.

Fig. 4. Ten-year incidence of CHD among clerical and nonclerical working women aged 45 to 64 years by marital status and children.

Fig. 5 Ten-year incidence of CHD among clerical and nonclerical working women aged 45 to 64 years by husband's occupation.

Fig. 6 Mean levels of the standard coronary risk factors among housewives, working women, and men aged 45 to 64 years at their eighth or ninth biennial medical examinations.

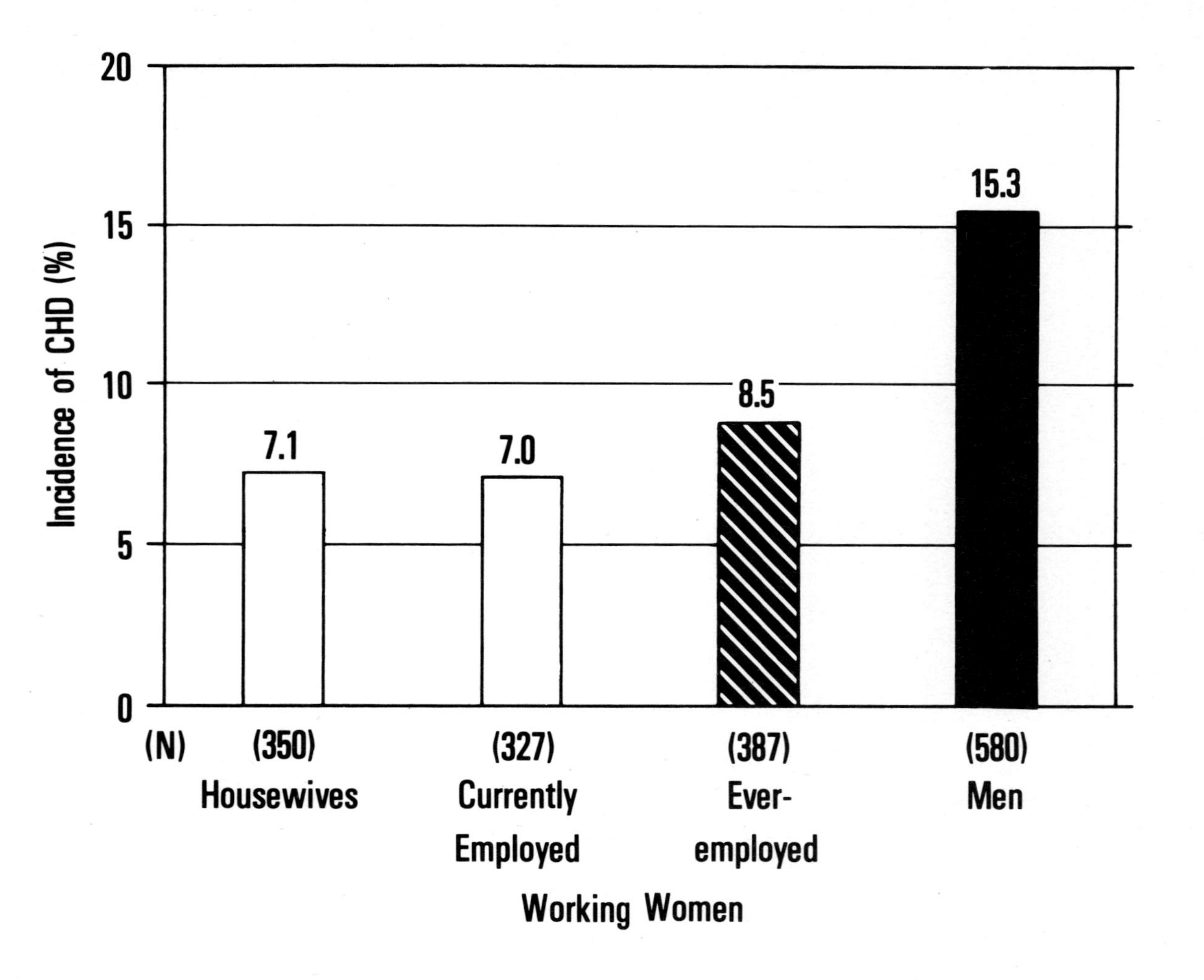
Incidence of CHD (%)
20
15
10
5
0
7.1
7.0
8.5
15.3
(N)
(350)
Housewives
(327)
Currently
Employed
(387)
Ever-
employed
Working Women
(580)
Men

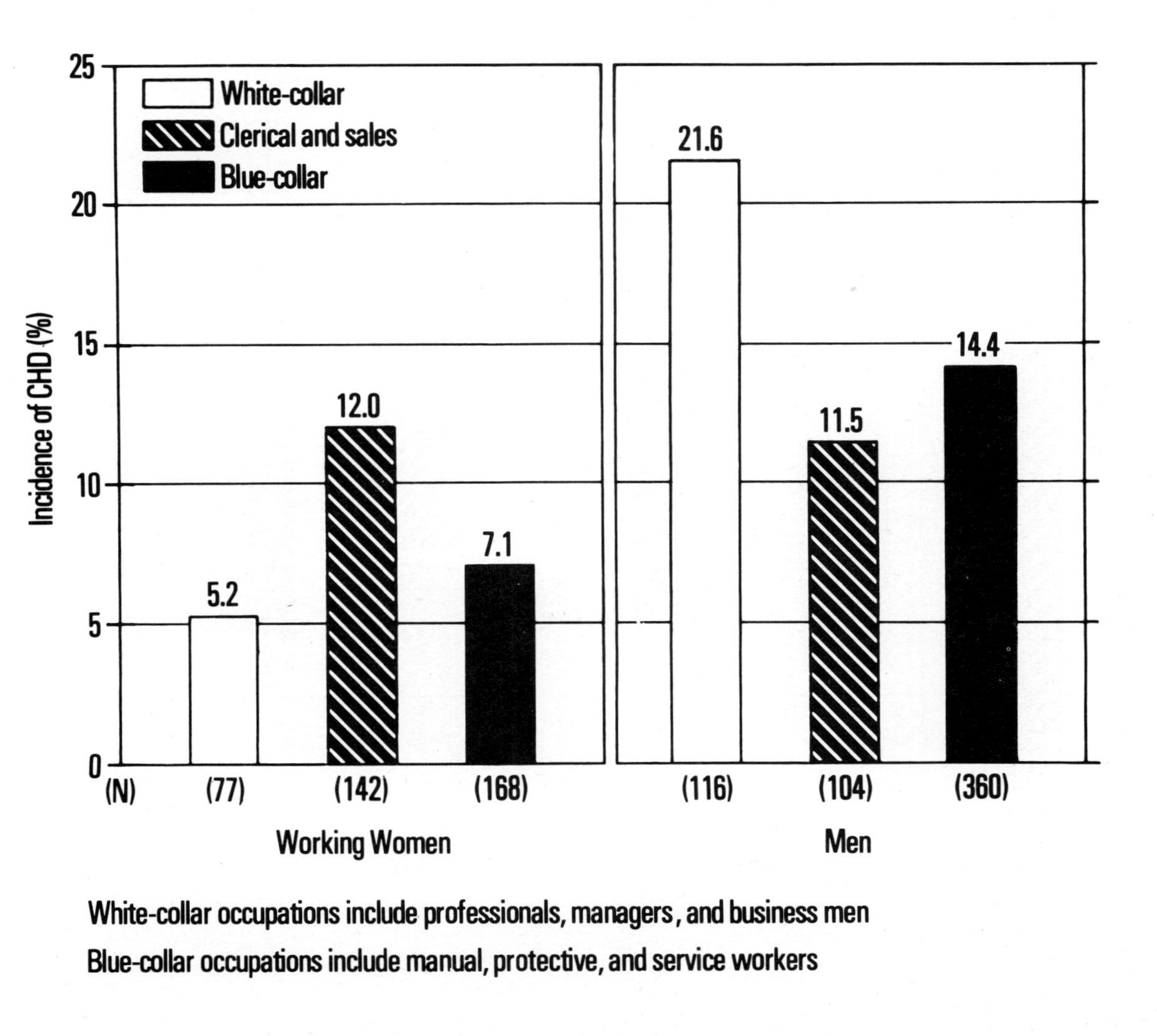

Incidence of CHD (%)
25
20
15
10
5
0
(N)
White-collar
Clerical and sales
Blue-collar
5.2
12.0
7.1
(77)
(142)
(168)
Working Women
21.6
11.5
14.4
(116)
(104)
(360)
Men
White-collar occupations include professionals, managers, and business men
Blue-collar occupations include manual, protective, and service workers

Incidence of CHD (%)
15
10
5
8.2
5.7
7.8
9.0
11.0
(N)
(171)
(159)
(77)
(146)
(91)
Number
of Children
1-2
3+
0
1-2
3+
Housewives
Working Women

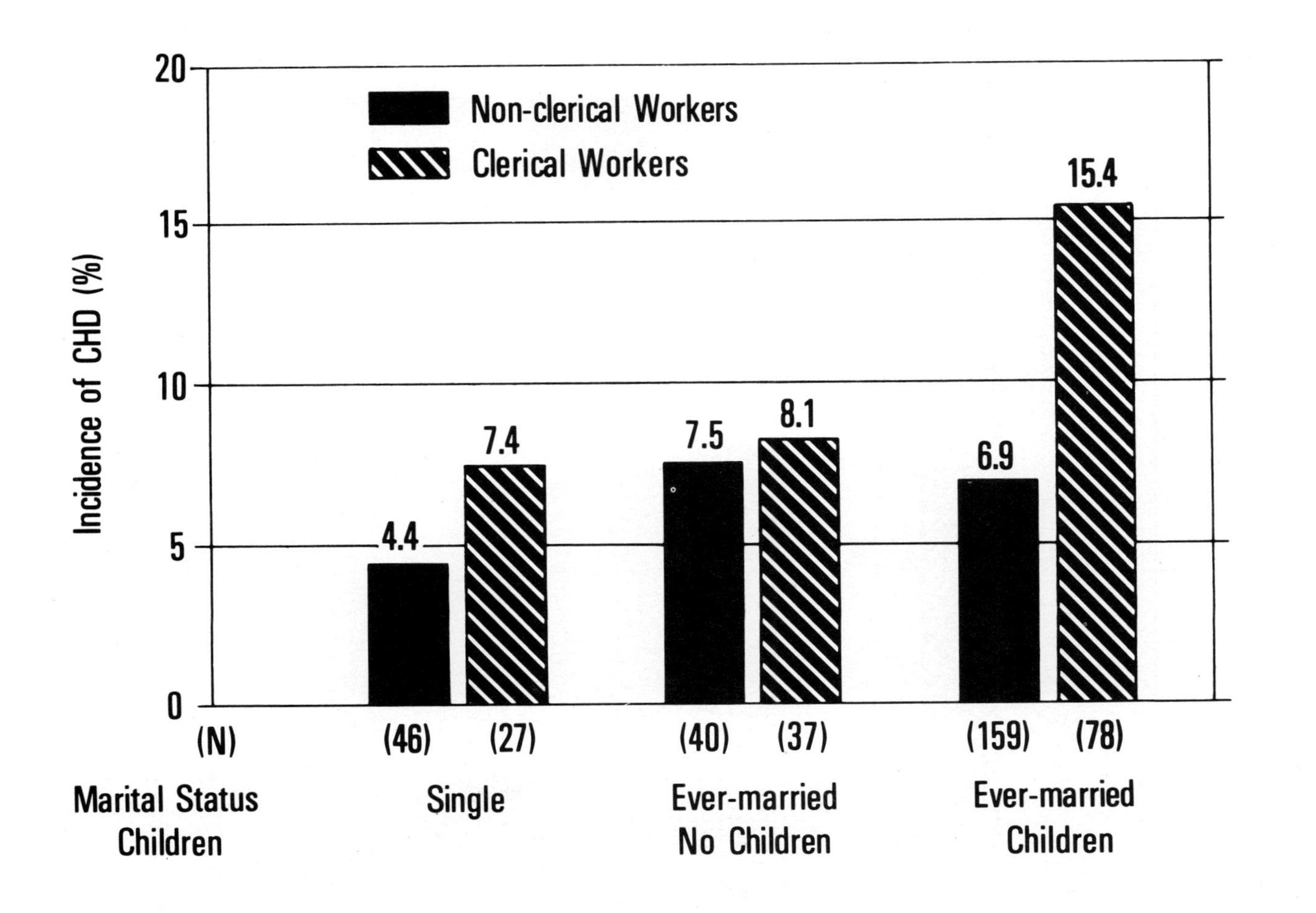

Non-clerical Workers
Clerical Workers
Incidence of CHD (%)
20
15
10
5
0
4.4
7.4
7.5
8.1
6.9
15.4
(N)
(46)
(27)
(40)
(37)
(159)
(78)
Marital Status
Children
Single
Ever-married
No Children
Ever-married
Children

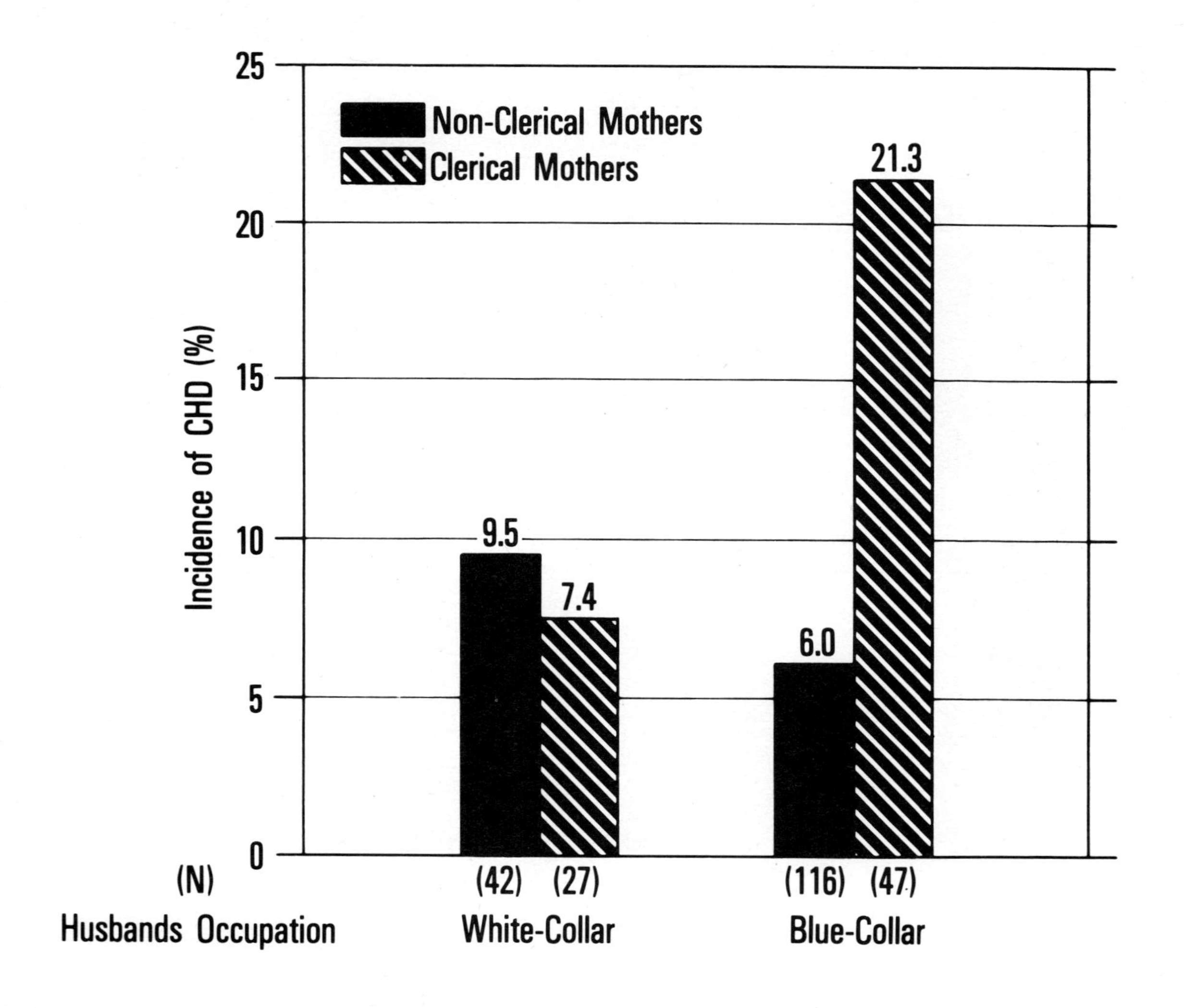

Non-Clerical Mothers
Clerical Mothers
Incidence of CHD (%)
25
20
15
10
5
0
21.3
6.0
9.5
7.4
(N)
(42)
(27)
(116)
(47)
Husbands Occupation
White-Collar
Blue-Collar

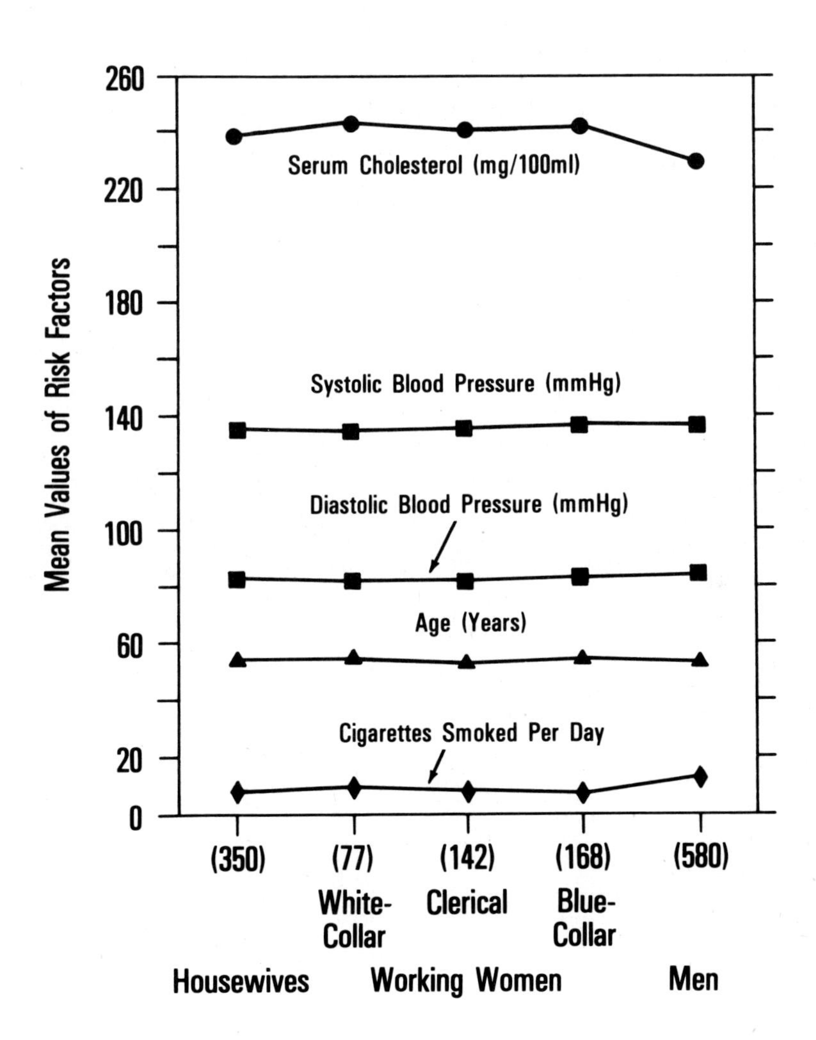

260
220
180
140
100
60
20
0
Mean Values of Risk Factors
Serum Cholesterol (mg/100ml)
Systolic Blood Pressure (mmHg)
Diastolic Blood Pressure (mmHg)
Age (Years)
Cigarettes Smoked Per Day
(350)
(77)
(142)
(168)
(580)
White-Collar
Clerical
Blue-Collar
Housewives
Working Women
Men

# SECTION 5
## Strategies for Alleviation Worksite Stress

# A COMPARISON OF WORKSITE RELAXATION METHODS

LAWRENCE R. MURPHY
Applied Psychology and Ergonomics Branch
Division of Biomedical and Behavioral Science
National Institute for Occupational Safety and Health
4676 Columbia Parkway
Cincinnati, Ohio 45226

## INTRODUCTION

Other chapters in this book present descriptions of job elements and work practices that create stress in clerical/secretarial workers. This chapter focuses on stress reduction and, specifically, the potential merits of work-based relaxation training programs. The exhaustive clinical literature on the effectiveness of relaxation methods for reducing arousal level and psychological signs of stress are not covered here nor is the literature regarding organizational approaches to stress reduction (for example, job redesign and job enrichment). Rather, this review will be limited to stress reduction approaches that are tailored to the needs of the individual worker and offered at the workplace.

## STRESS AND COPING

Stress is a common experience in our world. Feelings of tension, frustration, fatigue, or anger that surface readily during conflict situations at work and at home reveal the strain of coping with stress. Stress reactions are accompanied by physical and behavioral by-products that can include rapid heart beat, elevated blood pressure, muscle aches, stomach distress, social isolation, loss of appetite and incresed alcohol or tobacco use. These acute reactions can lead to more serious health problems such as hypertension, coronary heart disease, stomach ulcers, mental ill health, or alcoholism (ref. 1, 2, 3). The seriousness of these health problems is a function of the frequency and clustering of stressful events and the degree to which effective coping strategies are employed.

All of us have experienced stress at some time from childhood through adolescence to adult life. Unfortunately, many people have not had comparable experience in the art of stress reduction or stress coping. Organizations are similarly limited: they have far greater experience with stress than with effective strategies to alleviate stress and stressful work conditions. Indeed, in their recent review of literature in this area, Newman and Beehr (ref. 4) found virtually no evaluative studies of occupational stress reduction and forcefully make a case for filling this gap in the applied research literature. Since their review, several evaluative studies have assessed the benefits of stress management training in work settings. These studies have varied considerably in terms of methods used, program duration, number of sessions, and type of work group studied.

This chapter reviews studies that evaluate the merits of select strategies for managing work stress and discusses important issues in need of additional research.

## RESEARCH EVIDENCE

Stress management techniques that have been tested for their usefulness in work settings include meditation, progressive muscle relaxation, biofeedback, and cognitive/behavioral skills training. Each technique is briefly described below as a prelude to presenting research evidence for its utility in worksite programs.

### Meditation

A number of types of meditation have been promoted as a means of reducing stress. Transcendental meditation (TM) is perhaps the best known and involves the simple repetition of a secret mantra or prayer word while maintaining a passive mental attitude. (The mantra is a one-syllable word specifically chosen for each student). The procedure purports to produce physical relaxation and improve mental clarity although TM has not been scientifically evaluated for its usefulness in worksite stress reduction programs. However, a noncultic variation of TM, developed by Herbert Benson (ref. 5), in which the word "one" is repeated with each exhalation instead of a mantra or prayer word, has been used as a relaxation technique in two worksite studies.

Peters, Benson, and Porter (ref. 6) and Peters, Benson, and Peters (Ref. 7) taught 58 office workers the "one-technique" (ref. 5) while an additional 132 employees comprised three control groups: self-relaxation (N = 39), no instructions (N = 39) and nonvolunteers (N = 58). Employees were provided with educational information about stress and its health consequences and received one instruction session in the "one-technique." After an 8-week evaluation period (preceeded by a 4-week baseline), the relaxation response group showed the largest reductions in systolic (12 mmHg) and diastolic (8 mmHg) blood pressure and the most impressive improvements on questionnaire measures of symptom reporting, illness frequency, performance rating, and attitudes toward life and work. Of special significance here are the facts that participants were not selected for study because of high stress risk or the presence of stress symptoms and were _normotensive_ as a group at the start of the study.

In a follow-up evaluation of these participants 6 months later, Peters (ref. 8) found that the original benefits had been maintained but at reduced levels. Blood pressure levels at follow-up remained significantly below baseline levels, indicating the continued effectiveness of the stress management intervention.

A second study compared the "one-technique" with clinically standardized meditation (CSM) and progressive muscle relaxation in 154 telephone company employees who learned the techniques at home via instructional materials and cassette tapes (ref. 9). Clinically standardized meditation is a variation wherein the repetition of a word is not linked to each breath and attention is specifically directed toward a mental stimulus. At follow-up 5.5 months later, the meditation-based methods produced the largest decreases in reported depression, hostility, anxiety, and a general symptom severity index compared to the muscle relaxation or control group. The CSM group reported beneficial effects more frequently and saw the technique as facilitating social improvement more so than the remaining groups. Finally, CSM subjects reported practicing the method more frequently at follow-up compared to the remaining groups.

Taken together, these studies provide first indications of the benefits of meditation strategies for reducing physiological arousal levels and symptoms of stress. The lure of these techniques lies in their simplicity of instruction, short learning period, and absence of expensive equipment for implementation.

### Progressive muscle relaxation

The most frequently used strategy in worksite stress management programs has been progressive muscle relaxation (PMR). In this technique, individuals are instructed first to tense a muscle group and study the feelings of tension there and then to relax the muscles and notice how different the states of tension and relaxation are. By systematically moving through major muscle groups of the body, individuals become proficient at recognizing tension and creating a state of deep muscle relaxation (ref. 10).

As already noted, Carrington and others (ref. 9) found progressive muscle relaxation training to be more effective than controls, yet less effective than meditation methods in reducing psychological and somatic signs of stress. However, other authors have reported significant effects following PMR training. For example, Peterson (ref. 11) found that PMR training was more effective than cognitive/behavioral skills training in lowering blood pressure, heart rate, and forehead muscle activity while significantly raising hand temperature. Eighty-one Los Angeles county employees received 6 weekly 1-hour sessions in one of the above methods or a combination of the two following a one-week baseline period. PMR training was found to be the most effective method for reducing physiological arousal level and increasing subjective feelings of competence in coping with stress, although all three groups showed improvements on these measures compared to wait-list controls at follow-up 6 weeks later.

Murphy (ref. 12) compared PMR to biofeedback and self-relaxation techniques in nurses who received 6 daily training sessions following a 2-day baseline. PMR training resulted in the largest within-session decreases in forehead muscle activity immediately post-training and significant improvements at a 3-month follow-up in the ability to cope with stress and frequency of stress coping relative to self-relaxation controls. All groups in this study reported improvements in sleep behavior at follow-up. In a second study using highway maintenance workers, PMR participants showed EMG reductions similar to controls and smaller than the biofeedback group. On a variety of self-report measures, the PMR group and controls were not significantly different (ref. 13).

Schleifer (ref. 14) reported that cue-controlled relaxation (a variant of PMR) was more effective than self-relaxation in reducing forehead muscle activity levels after 6 weekly training sessions provided to retail store clerks. The cue-controlled group also reported greater decreases in trait anxiety relative to controls. On measures of blood pressure, state anxiety, and a stress management skills inventory, both experimental and control groups showed equivalent post training reductions. No follow-up data were reported.

Steinmetz, Kaplan, and Miller (ref. 15) taught 243 workers in mixed occupations a multimodal stress coping strategy that included PMR and various cognitive strategies. The program format varied in terms of the number of program days but typically involved 8 hours of training. Post-training evaluation indicated decreases in stress symptoms, perceived stress at work, and psychological reactions to stress. However, since individual responses to questionnaire items were not individually coded, the determination of program effects was done on a group basis, pre- versus post- training. Consequently, the results must be viewed with caution.

The above reports suggest that PMR can be an effective strategy for reducing arousal level as measured by forehead muscle activity and for alleviating psychological signs of stress such as anxiety. The usefulness of PMR for lowering blood pressure and increasing hand temperature is at best suggestive and requires additional supporting evidence.

## Biofeedback

In biofeedback, a person is provided with information or feedback about a psychophysiological function, and, over time, learns to exercise a degree of control over that function. For example, muscle electrical activity can be recorded and transformed into a tone whose pitch rises as tension increases and falls as tension decreases. With practice, individuals receiving such feedback come to learn which thoughts, feelings or images tend to raise the tone and which one tend to lower the tone. By concentrating on those that lower the tone, individuals can achieve significant reductions in muscle activity. Biofeedback training has been used successfully to lower heart rate, blood pressure, and other functions previously believed to be involuntary (ref. 16).

Manuso (ref. 17) selected 30 employees with a history of chronic anxiety or recurring headaches to receive stress management training three times per week over a 5-week period. This company-sponsored training program was multimodal and included the use of biofeedback, progressive muscle relaxation, breathing exercises, and imagery training. Outcome measures were forehead EMG, symptom reporting, clinic visits, and symptom interference with work, which were assessed in a 3-month period before and after stress management training. Manuso found improvements along _all_ dependent measures post-training relative to pre-training levels. While these results are impressive, a caveat is warranted since no control group was included in the study.

Manuso also computed cost/benefit ratios for the training program based upon the 30 participants. Using costs of clinic visits, time off the job to visit the clinic, work interference affecting bosses, co-workers, and subordinates, the author reported an impressive ratio of 1:5.52. That is to say, for each dollar spent in training, the corporation realized $5.52 in benefits from improved productivity and lower health care costs for each of the 30 participants. The expected 3-year savings to the corporation through reduced symptom activity of the 30 participants was $178,322.55--a sizeable return on investment.

As already noted, Murphy (ref. 12) utilized EMG biofeedback in a worksite study with nurses. No supplementary methods (for example, breathing exercises) were offered to biofeedback-trained nurses who received only 6 daily training sessions. Murphy reported significant increases in hand temperatire in the biofeedback group and a reduction in forehead EMG levels, although the latter was not statistically different from controls. Moreover, compared to either PMR or self-realization groups, biofeedback-trained nurses reported greater increases in work energy levels and perceived effectiveness of the coping strategy for reducing stress at work at follow-up 3 months later. A second effort with highway maintenance workers found that biofeedback-trained workers showed the largest decrease in EMG levels indicative of relaxation and the largest changes on subjective stress symptoms, although the groups were not statistically different on the latter measures (ref. 13).

These two studies suggest that biofeedback training can be effective in worksite training programs to reduce physiological arousal levels and symptoms of stress and improve workers' ability to cope with stress. The usefulness of biofeedback for reducng blood pressure has not been directly assessed in a worksite training program.

Cognitive/behavioral skills

Cognitive/behavioral (C/B) skills training represents a composite of several techniques designed to improve coping behavior and stress management. Components of such training may involve cognitive restructuring (which emphasizes the role of thoughts and perceptions in controlling feelings), assertiveness training (which teaches individuals to express feelings in honest and appropriate ways), and rational-emotive training (which deals with reducing negative self-talk).

Steinmetz and others (ref. 15) utilized cognitive and behavioral skills training along with PMR in an 8-hour stress management program for a variety of occupational groups. As already noted, the multimodal training program appeared to reduce the frequency of stress symptoms and decrease perceived stress at work, although questionnaires were not coded individually in this study. Since a combination training method was used, it is not possible to determine the unique effects of PMR and C/B skills training.

Peterson (ref. 11) compared the effectiveness of cognitive/behavioral skills training with PMR and a combined training group. Although PMR was judged most effective in reducing arousal level, C/B skills training was most effective in reducing trait anxiety and total number of stress symptoms reported. The group who were taught a combination of PMR and C/B skills training did not show additional benefits relative to each training method alone, perhaps as a result of the sheer amount of information presented in the short training program.

Finally, Scheingarten (personal communication, 1981) utilized a combination of PMR and C/B skills training in a 3-day seminar offered to eight steel company supervisors in Cincinnati, Ohio. Results indicated a significant decrease in both systolic (5 mmHg) and diastolic (5 mmHg) blood pressure immediately post-training, although the participants were normotensive at the start of the seminar. No control group was used and no follow-up was reported.

While these studies suggest a value for C/B skills training in worksite training programs, specific benefits attributable to this procedure cannot be clearly discerned. One reason for this state of affairs is that C/B skills training has been used along with PMR in two of the three studies reviewed here so that delineation of unique effects has not been possible. The exception is the study by Peterson (ref. 11), who demonstrated that C/B skills training does result in decreased anxiety and stress symptoms as well as reductions in physiological arousal level. The physiological arousal reductions were of smaller magnitude than those in the PMR group yet were statistically different from controls.

## FUTURE RESEARCH TOPICS

The field of stress management research is still very young and several important topics need to be addressed in future reasearch. Recognizing the distinct need for more demonstration studies, such efforts might additionally examine one or more of the following factors:

1. Skill Maintenance. Common to most health promotion programs, long term benefits are usually smaller than those observed immediately post training. There is a need to examine more closely factors that influence practice rate and manipulate strategies designed to foster more frequent use of learned health promotion skills.

   A first step in this direction has been taken by Peters (ref. 8), who reported higher practice rates in participants who taught the meditation/relaxation method to someone else and that practice rate in the first month post-training accurately predicted practice rate at 6 months, highlighting the first post-training month as a critical period for skills maintenance.

2. Blue- versus White-Collar Occupations. By and large, stress management programs have been offered to white- or pink-collar employees to the exclusion of blue-collar groups. The generalizibility of benefits to blue-collar occupations needs to be assessed. An on-going NIOSH study is being conducted with highway maintenance workers of the City of Cincinnati; preliminary results are quite impressive regarding the ability of participants to reduce EMG levels after relaxation training (ref. 13).

3. Generalization of Relaxation Skills. Worksite stress management studies have demonstrated significant reductions in physiological arousal level immediately post-training and, in one case, at a follow-up conducted 6 months later. No evidence, however, documents the usefulness of the acquired relaxation skills during actual stressful encounters. Although the relaxation skills can be effective as a preventive strategy lowering basal arousal levels, it would be interesting to examine effectiveness during acute stressful situations. In this regard, Kohn (ref. 18) has shown that individuals trained in progressive muscle relaxation made fewer performance errors under conditions of high stress compared to controls. This finding suggests that relaxation training may be useful in helping individuals cope with stressors that could impair performance. Additional research is necessary to replicate this finding and extend it to a variety of stress-producing situations and other relaxation strategies.

4. Cost-Benefit Analysis. In the final analysis, the decision of organizations to establish stress management programs is likely to be based upon monetary concerns. While it is important to conduct additional demonstration studies of the merits of work-based stress management programs, it will become increasingly important to provide indications of program benefits measured against costs. The difficulty in assigning monetary value to program benefits is not insurmountable; Manuso (ref. 17) has taken a first step in this direction and his analysis can serve as a prototype for others.

   In the ongoing NIOSH study with highway maintenance employees, the City of Cincinnati is conducting an independent evaluation of the training program using measures of absenteeism, minor traffic accidents, and performance ratings, among others. The resultant data can be used to compute return on investment figures for this program.

## CONCLUSION

While the number of evaluative studies in the area of worksite stress management are few, existing evidence suggests that such programs are effective. Benefits to employees in terms of reduced physiological arousal, reduced symptom activity, and reduced anxiety have been reported using diverse training strategies, (for example, biofeedback versus meditation) on various worker groups. However, too few studies have been conducted to determine the relative merits of each approach and weight cost concerns against obtained benefits.

Occupational stress management research will likely grow tremendously in the next few years should positive effects such as those indicated in this review continue to be found. The techniques have potential for improving the health and well-being of the worker and of offsetting the spiraling costs of occupational stress in terms of productivity losses and stress-related disorders.

Clearly, these methods do not represent a cure-all for job stress. Improving working conditions, job enrichment, and the application of sound ergonomic principles to job design are important tools for reducing work stress and represent primary, preferred strategies. Nevertheless, individual approaches may potentially improve the coping skills of workers and reduce negative signs of stress and can be integrated easily into existing employee assistance programs. Additionally, such training programs can be established quickly and evaluated for cost/benefit to the organization without major disruptions of the work routines or production schedules. Finally, stress management programs can be useful adjuncts to job redesign/organizational change approaches to teach workers greater control over stress reactions arising from both work and nonwork sources.

REFERENCES

1 C.L. Cooper and J. Marshall, Occupational sources of stress: a review of the literature relating to coronary heart disease and mental ill health. Journal of Occupational Psychology, 49, (1976), pp. 11-28.
2 R.A. Rose, C.D. Jenkins and M.W. Hurst, Air traffic controller health change study. Report #FAA-AM-78-39, National Technical Information Service, Springfield, Virginia, 1981.
3 H. Selye, Stress in Health and Disease, Butterworth Publishers Inc., London, 1976.
4 J.D. Newman and T. Beehr, Personal and organizational strategies for handling job stress: a review of research and opinion, Personnel Psychology, 32, (1979) pp. 1-43.
5 H.A. Benson, The Relaxation Response, William Morrow and Company, Inc., New York, 1976.
6 R. Peters, H. Benson and D. Porter, Daily relaxation response breaks in a working population: effects on self-reported measures of health performance and well-being. American Journal of Public Health, 67, (1977), 946-953.
7 R. Peters, H. Benson and J. Peters, Daily relaxation response breaks in a working population: effects on blood pressure, American Journal of Public Health, 67, (1977). pp. 954-959.
8 R.K. Peters, Daily relaxation response breaks in a working population: follow-up of a work-based stress reduction program, Report #PB-83-175364, National Technical Information Service, Springfield, Virginia, 1981.
9 P. Carrington, G.H. Collings, H. Benson, H. Robinson, L.W. Wood, P.M. Lehrer, R.L. Woolfolk and J.W. Cole. The use of meditation-relaxation techniques for the management of stress in a working population, Journal of Occupational Medicine, 22, (1980), pp. 221-231.
10 D.A. Bernstein and T.D. Borkovec, Progressive Relaxation Training, Research Press, Champaign, Illinois, 1973.
11 P. Peterson, Comparison of relaxation training, Cognitive restructuring/ behavioral training, and multimodal stress management training seminars in an occupational setting. Unpublished Doctoral Dissertation, Fuller Theological Seminary, 1981.
12 L.R. Murphy, A comparison of relaxation methods for reducing stress in nursing personnel, Human Factors, 25, (1983), pp. 431-440
13 L.R. Murphy, Stress management in highway maintenance workers, Journal of Occupational Medicine, in press.
14 L.M. Schleifer, Cue-controlled Relaxation: an innovative approach for reducing job stress in a hypertensive working population, Unpublished NIOSH report, 1981.
15 J. Steinmetz, R. M. Kaplan, and G. L. Miller, Stress management: an assessment questionnaire for evaluating interventions and comparing groups, Journal of Occupational Medicine, 24, (1981), pp. 923-931.
16 R.M. Lee, S.E. Baldwin and J.A. Lee, Clinical uses of biofeedback: a review of recent research, Henry Ford Hospital Medical Journal, 25, (1977), pp. 99-118.
17 J.S. Manuso, Executive Stress Management, The Personnel Administrator, November, 1979.
18 J.P. Kohn, Stress modification using progressive muscle relaxation, Professional Safety, 26, (1981), pp. 15-19.

# THE PROCESS OF RELAXATION TRAINING IN THE MANAGEMENT OF JOB STRESS

LAWRENCE M. SCHLEIFER, Ed. D.
Applied Psychology and Ergonomics Branch
Division of Biomedical and Behavioral Science
National Institute for Occupational Safety and Health
Centers for Disease Control
Public Health Service
Department of Health and Human Services

## INTRODUCTION

The premise that job stress may have adverse implications for worker health has prompted a growing number of employers in industry and government to establish stress-management training programs at the workplace. Employers increasingly recognize that stress-management programs may boost employee morale, increase productivity, and enhance the quality of working life. Stress-management training is viewed as a farsighted investment in the human resources of work organizations.

One of the more promising stress-management programs that has gained acceptance in occupational settings is relaxation training. Relaxation training refers to any set of procedures such as progressive muscle relaxation or meditation that evoke a psychophysiologic reaction called the relaxation response. See ref. 1, 2 for a more detailed description of these relaxation-induction techniques). The relaxation response consists of reductions in cognitive-emotional activity, blood pressure, heart rate, muscle tension and respiration rate (ref. 1). The therapeutic effects of relaxation training are based on the principle of reciprocal inhibition formulated by Joseph Wolpe (ref. 2): "If a response antagonistic to anxiety can be made to occur in the presence of anxiety-provoking stimuli so that it is accompanied by a complete or partial suppression of the anxiety responses, the bond between these stimuli and the anxiety responses will be weakened" (p. 17). Hence, job stress responses to adverse working conditions are counteracted or inhibited by relaxation responses.

Laboratory studies have demonstrated that relaxation techniques such as progressive relaxation and meditation can lower heart rate, blood pressure, respiratory rate, and subjective reports of distress (ref. 1, 3). In clinical settings, relaxation training has been successfully administered either as an adjunctive or definitive therapy in the treatment of a variety of stress-related disorders, including sexual dysfunction (ref. 4), public speaking anxiety (ref. 5), and insomnia (ref. 6).

More recently, relaxation training has been introduced in the workplace to prevent or alleviate stress reactions. One study found that relaxation training in an organizational setting reduced self-reported symptoms of stress (ref. 7). Another investigation, demonstrated that regular "relaxation breaks" at the workplace decreased blood pressure and had a positive influence on self-reports of health, performance and personal satisfaction among a group of office workers (ref. 8). Several stress-reduction studies conducted by the National Institute for Occupational Safety and Health (NIOSH) also found promising results with relaxation training among nursing personnel (ref. 9), highway maintenance workers (ref. 10), and hypertensive retail sales workers (ref. 11) (see ref. 12 for a more comprehensive review of this literature).

One of the more attractive features of relaxation training is that it can be employed to deal with behavioral, cognitive-emotional, and physiologic stress responses evoked by a wide range of job stressors, including work overload, role conflict and ambiguity, excessive work pace, shift work, and organizational change. In addition, it is relatively easy to learn how to elicit the relaxation response. Workers can learn to apply relaxation skills under stressful conditions in approximately six weeks of training and regular practice. Furthermore, training sessions can be provided on either an individual or group basis, allowing several workers to receive training at the same time. Finally, since the training sessions can be scheduled during lunch hours or extended breaks, there is minimal disruption of work-related activities.

Despite the apparent efficacy of relaxation techniques, occupational health professionals such as nurses, physicians, and psychologists may be unfamiliar with the procedural steps and practical considerations that are involved in the application of these techniques at the workplace. Relaxation techniques that have been designed for clinical applications must be modified when administered in occupational settings. Specific job stressors must be identified, their behavioral, emotional, and physiological effects evaluated, appropriate training procedures must be employed, and strategies for maintenance of long-term therapeutic benefits must be developed. The purpose of this chapter is to provide some direction to the occupational health professional in how to implement the process of relaxation training. It will go beyond the specifics of any given relaxation technique such as progressive muscle relaxation or meditation, and instead focus on the procedural steps that are so critical to the efficacy of relaxation training.

## ASSESSING THE NEED FOR RELAXATION TRAINING

The relative ease of administration and apparent effectiveness of relaxation training does not warrant its indiscriminate use with all job stress problems. Since relaxation training is an individual, as opposed to an organizational approach to stress management, it is not a panacea for job stress. It enables the worker to obtain symptomatic relief by modifying or alleviating stress reactions. It does not alter the excessive task demands or negative ergonomic factors that induce the stress response. In this regard, relaxation training is a stress-management technique of last resort. It should not be employed to adapt the worker to an unfavorable work environment but rather to alleviate stress that is otherwise unavoidable. In summary, relaxation training should not be administered until more direct interventions such as job redesign or organizational change have been considered.

One approach for determining the need for stress management is to have employees monitor their stress levels. Self-monitoring (ref. 13) is an effective information gathering technique that can be used to determine whether relaxation training is appropriate for dealing with a particular job stress problem. Self-monitoring involves the systematic observation of job stress reactions by the worker. The worker is asked to keep a daily log of the date, time, location, significant others present, a description of the stress-provoking conditions, and the nature and magnitude of stress-reactions.

An illustrative example of sample log entries can be found in Table I. Inspection of the entries reveals that the worker was under considerable deadline pressure to type up a thirty-page manuscript. Under these unfavorable conditions, the worker experienced a range of stress reactions, including tension, anger, fatigue, tachycardia, loss of appetite, and insomnia. Of note is the fact that the worker's stress reactions persisted even after the workday ended and interfered with daily eating and sleeping patterns.

Self-monitoring provides valuable information about the causes and consequences of job stress reactions. This information can be used to determine whether relaxation training or alternative stress-management interventions are indicated. For example, the worker experiencing deadline pressure might be encouraged to talk with her supervisor about the possibility of scheduling work activities so as to avoid this kind of problem. However, if such a solution is not possible because the supervisor has no control over the setting of deadlines, then a stress-management procedure such as relaxation training would be indicated. In summary, the illustrative example in Table I underscores the need for a systematic assessment of stress reactions prior to administering relaxation training.

Transfer of Relaxation Training Effects

An important goal of relaxation training is to transfer the effects of training to the occupational setting. While it is often assumed that the relaxation response will generalize automatically to stress-provoking conditions in the work environment, a number of clinical studies suggest that there may be incomplete transfer of training effects. For example, it has been observed by Chang-Liang and Denney (ref. 14) that in studies which did not encourage application practice of the relaxation response under stress-provoking conditions the relaxation methods were largely unsuccessful in reducing tension (ref. 15, 16, 17, 18, 19). However, relaxation methods that included application practice under stress-provoking conditions were successful in alleviating stress (ref. 20, 21, 22, 23, 24). Overall, the results of these studies support the need for application practice of the relaxation response in stress-provoking situations.

Prior to initiating application practice in the work setting, the relaxation response should be practiced in a less threatening environment using systematic desensitization, a procedural variation of relaxation training which facilitates application practice under simulated conditions. Systematic desensitization was developed by Joseph Wople (ref. 1) and is considered to be one of the most effective clinical procedures for controlling stress-related disorders. In fact, the exhaustive research which supports the effectiveness of systematic desensitization prompted Paul (ref. 25) to state "that systematic desensitization is the first psychotherapeutic procedure in history to withstand rigorous evaluation" (p. 146).

The first step in the desensitization process is to train the worker to achieve a state of deep relaxation. While any relaxation induction technique can be used (e.g., meditation, biofeedback, etc.), Wople (ref. 1) recommends progressive muscle relaxation. This involves alternately tensing and relaxing the various muscles groups of the body until a state of deep muscle relaxation is achieved (see ref. 26, 27). With practice, the relaxation response can be elicited by simply recalling the physical sensations associated with relaxation.

The next step involves the identification and categorization of job stressors according to different stress-provoking themes (e.g., deadline pressure or work overload). The stimuli or scenes comprising each theme are then ranked on a scale from 1-10 according to their relative stress levels. The self-monitoring assessment procedure described earlier provides the "raw data" for constructing the stress hierarchies. Examples of different stress hierarchies can be found in Table II.

The final step is desensitization proper. While in a relaxed state, the worker is asked to imagine the least stressful item on the hierarchy and work gradually toward the most stress provoking. If any stress reactions are experienced while imagining a given scene, the worker is asked to stop imagining the scene and attempt to regain the state of deep relaxation. Presentation of the scene is repeated until the worker can imagine the stress-provoking situation without experiencing excessive discomfort.

## THE USE OF APPLICATION TRAINING

Upon completion of the desensitization process in a nonthreatening environment, application practice of the relaxation response should be initiated under stress-provoking conditions in the work setting. A relaxation training procedure that facilitates application practice under such conditions is cue-controlled relaxation. Cue-controlled relaxation is based upon a classical conditioning model for inhibiting the stress response. The relaxation response is conditioned to a self-produced word (e.g., relax). With practice, the cue word exerts stimulus control over the relaxation response.

Paul (ref. 28) describes the cue-controlled relaxation procedure as follows:

> After the client is totally relaxed, he is instructed to focus all of his attention on his own breathing and then to subvocalize a cue word each time he exhales--such as "calm" or "relax." The therapist repeats the word in synchrony with exhalation 5 times and the client continues for 15 more pairings. After repeating this procedure over a period of 4 to 5 weeks, with the client giving 20 additional pairings on his own each night following relaxation practice, the ability of the self-produced cue word to bring about relaxation may be tested in the office. This is done by having the client imagine a threatening situation until some degree of anxiety is experienced, then he is instructed to take a deep breath, and subvocalize the cue word anytime he begins to feel a slight inappropriate increase in any real-life situation (ref. 28).

Cue-controlled relaxation is an effective stress-management technique that has been successfully employed in the clinical treatment of test anxiety (ref. 29, 30), flight phobia (ref. 31) and social-evaluative anxiety (ref. 32). It also was found to be effective in reducing muscle tension under laboratory-induced stress conditions (ref. 33). Moreover, cue-controlled relaxation appears to be a promising technique for controlling job stress (ref. 11). While it has been employed primarily with progressive muscle relaxation, it is possible to adapt cue-controlled relaxation to any technique that elicits the relaxation response.

In addition to promoting the transfer of training effects, application practice shifts the locus of therapeutic control from the occupational health practitioner to the worker. Instead of being constrained to a relatively passive role of faithfully following the practitioner's relaxation instructions in the training setting, the worker elicits the relaxation response under stress-provoking conditions in the workplace. Hence, the worker is more likely to attribute beneficial effects of training to the newly-acquired relaxation skills, and not the practitioner's unique personality or other nonspecific factors.

Application practice also diminishes the worker's lack of control over unfavorable work conditions. This is particularly important in light of the recent research which indicates that a lack of control over the conditions of work increases the risk for coronary heart disease (ref. 34). While it may not always be possible to modify adverse job situations, at least the response to such conditions can be modified or self-regulated by the worker. The exercise of such control may be highly therapeutic.

## MAINTENANCE OF TREATMENT EFFECTS

Despite the obvious importance of sustaining the therapeutic effects over time, it is not uncommon for trainers to simply terminate training without any attempt to promote long-term maintenance. This often occurs as a result of the mistaken notion that relaxation can provide permanent relief from job stress. However, as previously indicated, relaxation training is not a panacea for job stress. Unless the aversive conditions of work are modified, there probably will be a reoccurrence of symptoms. In fact, such a possibility underscores the need for the development of maintenance strategies.

One approach to promoting maintenance is to encourage the worker, upon completion of training, to elicit the relaxation response in stress-provoking situations on a regular basis. Frequent application practice diminishes the potency of job stressors and minimizes symptom recurrence. In most cases, the alleviation of stress symptoms usually provides sufficient motivation to engage in application practice. However, the worker may, in some instances, find it difficult to adhere to a regimen of regular application practice. In order to minimize such a possibility, the trainer should hold booster sessions that reinforce regular application practice of relaxation skills. Another approach to promoting maintenance is to use the logs described earlier to monitor application practice. The logs could be reviewed periodically to determine the frequency of stress-reactions and the need for application practice. Moreover, the logs would provide a systematic means of charting the progress of relaxation training in alleviating stress-reactions.

The presentation of the therapeutic rationale is another important factor in promoting maintenance. A therapeutic rationale that has excessive philosophical or religious overtones, or over emphasizes the unique style or personality of the therapist, increases the possibility that the worker will attribute the effects of training to the therapist or other vaguely-defined factors, and not the newly-acquired relaxation skills. This type of rationale shifts the locus of therapeutic control away from the workers and diminishes the potential for long-term maintenance of relaxation skills. Instead, the therapeutic rationale should emphasize that relaxation training is based upon a set of scientifically derived procedures which enable the worker to self-regulate or modify reactions to unfavorable working conditions.

A final strategy for promoting maintenance is to encourage workers who have acquired relaxation skills to train their peers in how to elicit the relaxation response. Workers who are willing to take the time and effort to train co-workers are more likely to utilize their relaxation skills on a regular basis. Moreover, workers who are also trainers serve as excellent role models and can be very effective in encouraging co-workers to participate in a relaxation training program. With such an approach the prospects for long-term maintenance of relaxation effects are enhanced considerably.

## CONCLUSION

Relaxation training appears to be an effective method for dealing with job stress. There are few other stress-management approaches that have application to such a wide range of behavioral, cognitive-emotional, and physiologic stress reactions. However, relaxation training is not a panacea for controlling job stress. Its role should be limited to dealing with stress reactions that are not amenable to more direct interventions such as job redesign or organizational change.

Occupational health professionsls must avoid the tendency to administer relaxation techniques in a mechanical "cookbook" fashion. There needs to be more consideration given to other aspects of the training process. In this regard, the need for relaxation as opposed to alternative interventions must be demonstrated, the effects of relaxation must be transferred from the training to the occupational setting, and the long-term maintenance of relaxation effects must be ensured. In summary, these procedural steps are central to the relaxation training process.

## REFERENCES


1 H. Benson, The Relaxation Response, William Morrow, New York, 1975.

2 J. Wolpe, The Practice of Behavior Therapy, Pergamon Press, New York, 1973.

3 G. L. Paul and R. W. Trimble, Recorded versus "live" relaxation training and hypnotic suggestion: Comparative effectiveness for reducing physiological arousal and inhibiting stress response. Behavior Therapy (1970) 285-302.

4 S. Asirdas and H. R. Beech, The behavioral treatment of sexual inadequacy. Journal of Psychosomatic Research, 19 (1975) 345-353.

5 G. L. Paul, Insight Versus Desensitization in Psychotherapy. Stanford University Press, Stanford, 1966.

6 M. Kahn, B. L. Baker, J. Weiss, Treatment of insomnia by relaxation training. Journal of Abnormal Psychology, 23, (1968), 556-558.

7 P. Carrington, G. Collings and H. Benson, The use of meditation relaxation techniques for the management of stress in a working population. Journal of Occupational Medicine, 22, (1980), 221-231.

8 R. K. Peters, H. Benson and J. M. Peters, Daily relaxation response breaks in a working population. American Journal of Public Health, 67, (1977), 946-959.

9 L. R. Murphy, A comparison of relaxation methods for reducing stress in nursing personnel. Human Factors, 25, (1983), 431-440.

10 L. R. Murphy, Stress management in highway maintenance workers. Submitted for publication.

11 L. Schleifer, Cue-controlled relaxation: An innovative method for reducing job stress in a hypertensive working population. Submitted for publication.
12 L. R. Murphy, Occupational stress management: A review and appraisal. Journal of Occupational Psychology, in press.
13 F. H. Kanfer. Self-management methods. In F. H. Kanfer and A. P. Goldstein (Eds.) Helping People Change, Pergamon Press, New York, 1975.
14 R. Chang-Liang and D. Denny, Applied relaxation as training in self-control. Journal of Counseling Psychology, 23, (1976) 183-189.
15 J. F. Aponte and C. E. Aponte, Group pre-programmed systematic desensitization without the simultaneous presentation of aversive scenes with relaxation training, Behavior Research and Therapy, 9 (1971) 337-346.
16 G. Cooke, Evaluation of the efficacy of the components of reciprocal inhibition psychotherapy. Journal of Abnormal Psychology, 73, (1968) 464-467.
17 G. C. Davison, Systematic desensitization as a counter-conditioning process. Journal of Abnormal Psychology, 73, (1968), 91-99.
18 S. Rachman, Studies in desensitization: I. The separate effects of relaxation and desensitization. Behavior Research and Therapy, 3, (1965).
19 D. Rimm and D. Medeiros, The role of muscle relaxation in participant modeling. Behavior Research and Therapy, 8, (1970), 127-132.
20 D. R. Denney, Active, passive and vicarious desensitization. Journal of Counseling Psychology, 21, (1974) 369-375.
21 C. Folkins, K. Lawson, E. Option and R. Lazarus, Desensitization and the experimental reduction of threat. Journal of Abnormal Psychology, 73, (1968) 112-118.
22 N. W. Freeling and K. M. Shemberg, The alleviation of test anxiety by systematic desensitization. Behavior Research and Therapy, 8, (1970) 293-299.
23 R. M. Laxer and K. Walker, Counter-conditioning versus relaxation in the desensitization of test anxiety. Journal of Counseling Psychology, 17, (1970), 431-436.
24 M. D. Spiegler, R. M. Liebvert, M. J. McMains and C. E. Fernandez, Experimental development of a modeling treatment to extinguish persistent avoidant behavior. In R. D. Rubin and C. M. Franks (Ed.), Advances in Behavior Therapy 1968, Academic Press, New York, 1979.
25 G. L. Paul in C. M. Franks (Ed.), Outcome of Desensitization. Behavior therapy: Appraisal and Status, 1969.
26 E. Jacobson, Progressive Relaxation, (University of Chicago Press, Chicago, 1938.
27 D. A. Bernstein and T. D. Borkovec, Progressive relaxation training: A manual for therapists, Pergamon Press, New York, 1973.
28 G. L. Paul, The specific control of anxiety, paper presented at the 1966 American Psychological Association Symposium.
29 R. Russell and J. Sipich, Cue-controlled relaxation in the treatment of test anxiety. Journal of Behavior Therapy and Experimental Psychiatry, 4, (1973) 47-49.
30 R. Russell, D. Miller and L. June, A comparison between group systematic desensitization and cue-controlled relaxation in the treatment of test anxiety. Behavior Therapy, 6, (1975) 172-177.
31 J. Reeves and W. Mealiea, Biofeedback-assisted cue-controlled relaxation for the treatment of flight phobia. Journal of Behavior Therapy and Experimental Psychiatry, 6 (1975) 105-109.
32 L. Schleifer, The effectiveness of group administered cue-controlled relaxation in the reduction of public speaking anxiety. Unpublished doctoral dissertation, 1978.
33 J. W. Ewing and H. H. Hughes, Cue-controlled relaxation: Its effects on EMG levels during aversive simulation, Journal of Behavior Therapy and Experimental Psychiatry, 9 (1978) 39-44.
34 R. A. Karasek, D. Baker, F. Marxer, A. Ahlbom and T. Theorell, Job decision latitude, job demands, and cardiovascular disease: A prospective study of Swedish men. American Journal for Public Health, 71, (1981), 696-704.

Table I

Occupational Stress Assessment Log

| Date | Time | Location | Significant Others Present | Description of Stress Provoking Situation | Description of Stress Response | Magnitude of Stress Response on a scale from 1 to 10 |
|---|---|---|---|---|---|---|
| 7-27 | 1:00 pm | At my work station | Supervisor | My supervisor informed me that I had to type a 30-page manuscript by 5:00 p.m. | My stomach tightened up like a knot; I became angry | 8 |
| 7-27 | 2:15 pm | At my work station | | Typing the manuscript | My heart was pounding and I was feeling very tense | 8 |
| 7-27 | 3:45 pm | At my work station | Supervisor | My supervisor asked me if I would be finished by 5:00 pm; I told him that I would do my best to get it done by 5:00 pm | I kept on thinking about what my supervisor would do to me if I didn't complete the manuscript | 8 |
| 7-27 | 5:00 pm | At my work station | Supervisor | I gave my supervisor the typed manuscript; I didn't get even a word of thanks for getting the manuscript done on time | I was furious with my supervisor; I also felt very tired. | 8 |
| 7-27 | 6:00 pm | At the | Family dinner table | Thinking about my stressful day at the office | I didn't eat my dinner | 7 |
| 7-27 | 1:00 am | Sitting in kitchen | Husband | I was telling my husband how much I disliked my supervisor for giving me so much work at the last minute | I couldn't sleep; anger | 7 |

TABLE II
Examples of Stress Hierarchies

Deadline Pressure

1. A week before a deadline
2. Three days before a deadline
3. Two days before a deadline
4. On the day before a deadline
5. On the day of the deadline
6. The supervisor informs the worker of the deadline
7. The worker begins the task that must be completed before the deadline
8. The supervisor asks the worker whether the deadline will be met

Argument with Supervisor

1. Worker discusses with supervisor the best method for accomplishing a difficult task
2. Supervisor disagrees with worker
3. Worker defends his/her position
4. Supervisor begins to shout at worker
5. Supervisor insists that he/she is always right and storms out of the office

# JOB STRESS--LABOR AND MANAGEMENT ISSUES

EUGENE V. MARTIN
Consultant
6812 6th Street, N.W.
Washington, D.C. 20012

Two objectives provide the motivation for this presentation: first, to report on an action project on job stress in the graphic arts industry; second, to share a concern about our understanding of sex-role socialization and its possible effect on stress research.

The Job Stress Project grew out of mounting interest and concern in the Graphic Arts International Union (GAIU) with the occupational stress that confronts more than one million men and women who work in the industry. Where and how do you begin to plan and implement a comprehensive, industry-wide effort? What programs will be most relevant to workers? What will increase awareness of the issues and commitment to action among management and union leaders? What help can we offer individuals beyond relaxation techniques and sermons on lifestyles?

Because our resources are very limited, we could hardly expect to answer these questions, but we could test some initial steps using qualitative evaluation methods to derive conclusions that would inform subsequent efforts. Our activities can be considered as an example of a way to use the action research model as a paradigm for planned change. Our primary focus was on the hypothesis that joint labor-management action on job stress is initially appropriate and practicable.

## PROJECT BACKGROUND

In 1979, the GAIU initiated the Safety and Health Awareness and Action Program for Employees and Employers (SHAPE). GAIU's primary objective was to develop an educational program to be presented by local unions in conjunction with employers. The course seeks to increase individual awareness of occupational safety and health issues specific to the graphic arts industry. Moreover, it is intended to promote collective action with the maximum feasible degree of labor-management cooperation in order to protect and improve workers' lives.

SHAPE's first public activity was to conduct a survey of the interests and concerns about safety and health expressed by a sample of workers, local union leaders, and employer representatives. Respondents identified "mental health in the workplace" as a major concern. Accordingly, when the initial SHAPE training course was tested in use, an entire 3-hour session of the course was devoted to the topic. (The full SHAPE course now consists of 8 sessions.) The participants in the course were workers, local union officers, and managers. On both sites for testing the course, Washington, D.C. and Detroit, Michigan, the participants expressed a high priority for having more time and attention focused on job stress, with particular emphasis on the graphic arts industry.

We developed the Job Stress Project in response to this demand. At the time, the issue of job stress was new for both labor and management decision-makers. Neither the GAIU nor the Printing Industry Association (PIA) had adopted policy statements or undertaken action specifically addressing job stress as an industry-wide problem. (There are, of course, scattered efforts--such as employee assistance programs--that have been initiated by various local unions and/or employers.)

## PROJECT ACTIVITIES

Our goal was to promote appropriate action for dealing with job stress by labor and management. Our specific objectives were to assess the job stress situation in the industry as perceived by workers, union and management; identify the feasibility of labor-management cooperation in dealing with job stress; provide policy-makers in the industry--labor and management--with options for action; and improve our ability to help people deal with the specific job stressors in the industry.

Our hypothesis was that the people responsible for action needed answers to two basic questions: what is the job stress situation, and what can be done about it?

We designed the project as action research, an on-going replication of cycles of inquiry, planning and action. The major project activities were two related processes; we interviewed local and national policy-makers and we conducted a sequence of experimental workshops.

### Policy-maker interviews

In both Detroit and Washington we formed advisory panels of influential labor and management leaders; we also formed a third advisory group of national-level policy-makers. Our concern was to obtain support from the formal and informal power structure of the local or national portion of the industry. In general, we did not know their levels of awareness of job stress nor what views they held on the subject.

The individual interviews of each of the advisors concerned three aspects of job stress in the graphic arts industry. First, what is the current situation? (Do you think job stress is a significant issue? What produces it? What are labor and management's respective concerns?) Second, what is the desirable situation? (What long-range goals do you think are appropriate? What next steps would be useful? What will affect labor-management cooperation?) and third, what issues should be explored? (What should this project investigate? What do you personally want to learn about?)

Local Workshops

The advisory panels in Detroit and Washington nominated the participants for the workshops. The advisors could participate in, observe, or ignore the workshops. (The majority of them participated as fully as their schedules permitted.) We recruited some 15 to 20 nominated participants in each city to obtain a 5-way cross-section of the industry. We wanted equal numbers of workers from each of the three segments represented by GAIU; the preparatory, press, and finishing/binding work areas. Furthermore, we wanted comparable numbers each of local union leaders and management.

There were 6 workshops in each city, designed as a sequence that would both enable participants to learn about stress and help us identify worker-relevant ways of dealing with job stress. Workshop I was a 3-hour evening session explaining the program and reviewing the basic concepts of stress and stress dynamics. Workshop II, the following evening, set participants working to help each other develop a plan for assessing the stress situation in their individual worksites. Participants carried out their individual assessment plans during the following 2 weeks. Workshop III, held on a Friday night, was a report by participants on the results they obtained. Workshop IV, the next day, provided participants with a sampling of a number of approaches for dealing with stress, including ways to mediate stress individually, the use of support networks, and techniques for collective action to change external conditions.

During the following 10 to 12 days, participants considered or tested these approaches in their workplaces. Workshop V was a 3-hour evening meeting in which participants summarized and discussed their observations on the usefulness and appropriateness of correction measures. Workshop VI, the following evening, focused primarily on summarizing participants' personal ideas and the recommendations for action addressed to the industry's leadership and policy-makers.

The final step of the project was a report for the policy-makers on the products of the interviews and workshops. We originally intended to re-interview these policy-makers after they had considered their peers' and constituents' views. On the basis of their reactions, we had hoped to develop long-range goals and to identify pilot improvement efforts with volunteering employers and GAIU locals. As this is written, the current administration has ruled out Federal funding for social science research and further National Institute of Mental Health (NIMH) support for the project appears unlikely. GAIU officials are exploring other ways and means to continue this work.

PROJECT FINDINGS

On the basis of opinions expressed by experts who are policy-makers in the industry, and supported by the perceptions of shop workers and supervisors, we drew the following conclusions:

Almost without exception, workers and policy-makers recognized and described widespread distress--in the shop and in the office--for workers, for managers, and for the union leaders. Existing high job stress levels are damaging to individuals and destructive to productivity. Although the role and significance of any one stressor varied from one part of the industry to another, there was no segment or level in the hierarchy that appeared to be less at risk than another.

The most anxiety-provoking stressors for workers centered on their treatment as persons: (1) management that ignores their skill, their commitment, and their potential as a valuable source of ideas and information for improving operations; (2) supervisors whose styles are routinely critical, coercive, suspicious, inconsistent, unfair, and virtually never appreciative, adult, candid nor open to reason; (3) job assignment procedures that never recognize the individual for skill or accomplishment; and (4) work rules that appear arbitrary, inflexible, and demanding.

The list of frequently identified stressors for workers included environmental factors such as high heat and noise levels, toxic substances and concerns about chemicals, and seemingly unending increases in machine speeds/production rates and automation of equipment and processes.

Although many of the management representatives described the tension of decision-making as bearing personal responsibility for the work as a source of stress for managers only, there was no difference in the way workers and managers described these concerns. Indeed, the more skilled workers often reported having to fight supervisory interference and quantity demands in order to provide work of sufficient quality to meet customer expectations. But even the less skilled workers often expressed high levels of personal concern, challenge, and involvement with their jobs. They were intensely aware that production levels have a direct effect on their livelihoods and asserted that management is unaware of the skill actually required not only to do the routine job but to maintain a stupefying process successfully for long periods of time.

Action to improve the job stress situation is viewed as an acceptable area for labor-management cooperation. The range of opinions expressed in interviews varied from disbelief and opposition to cautious optimism. The consensus was that there is a substantial area of mutual benefit and that cooperation is the preferred way to begin to deal with the issue. In addition, there is a clear sense that eventually some issues will have to be confronted at the bargaining table. (It should also be noted that management involved in this project represented employers whose employees are organized.)

The workshop design was constructed as an inquiry into workplace conditions as well as an educational program for the participants. The design maintained a primary focus on the external, workplace conditions; as we hoped, this general context did permit and support personal exploration. In fact, participants frequently discussed the issues in intensely personal terms. In Workshop VI, both groups expressed the unanimous, strong recommendation that further educational efforts build on experimentation with this design and its emphasis on using participant-centered discussion techniques for adults. The interview process with policy-makers appears to have substantially increased their awareness of the problem and their estimate of its significance to the industry and relevance to their concerns, besides providing data about their current positions, information needs, and decision criteria on job stress issues.

## SEX ROLE SOCIALIZATION AND JOB STRESS

This study has led to speculation about the possibility that the traditional concepts of job stress and of masculinity may lead most men--and some women--to mask and reinterpret their experience of stress in a manner that will significantly affect our efforts to study job stress.

As I conducted the interviews of the Job Stress Project, I became aware of a pervasive, general discomfort about dealing with job stress. Most of the men I talked with said early in the interview that they were willing to discuss job stress with me but they doubted that many others would be. Despite the nearly unanimous agreement among those interviewed that job stress is a critical problem in the industry, few imagined that others shared or would admit to similar views. A few people took strong exception to the interview: a management representative told me "stress cannot be studied scientifically" and a labor leader objected to the project as insulting to workers: "Our people aren't crazy." One union staff member initially denied that there was any significant stress in the industry and, later in the interview, paused abruptly, thought quietly for a while, and then said, "You know, as we've been talking I've been remembering the people I worked with when I was a shop steward. As I see their faces I remember the specific problems and concerns they had. I can't believe I told you there was no stress."

The manager of a plant then operating 7 days a week, 3 shifts a day, also denied at the beginning of the interview that job stress was a problem in his shop, "We've got good working relations here." Later in the interview, he told me that he occasionally--maybe two, three times a week--finds it "helpful" to go out to the plant roof where he can hold onto a chimney and scream "to relieve some tension."

I believe that many policy-makers think of stress as dealing with a "personal issue," "just emotions and feelings," "mental problems" that are inappropriate, "soft" concerns, not really suitable for public discussion. A tough manager, or a good worker--in this view--should be able to take it, to deal with stress without faltering or complaining. I share the reaction of a workshop participant who said, "The more we get into this, the more surprised I am that we've never paid any attention to stress problems before, never discussed it at a union meeting--or among the guys in the shop."

I also noticed a language pattern that related work to war, another traditional proving ground for masculinity. In one interview, I was told "Going to work is like being in a war movie--you know, where they say 'Nothing matters, just get the job done.'" A woman worker, telling me why some women might not want a higher paying man's job, said "You have to get there every day, be on time, and stay until the last shot is fired." In our work on stress we seem to be encountering a significant aspect of the way that men are socialized. The concept of a "hard worker" is virtually the same as the concept of a "real man." And the sex-role socialization of men _is_ to be workers--or warriors. These social patterns create the expectations that women as well as men have of themselves and of others in the workplace.

We have begun to examine the ways in which women are socialized. Traditional concepts of femininity are often tragically unjust to the individual and dysfunctional for our society. But we have not yet discussed as fully the way in which men are limited and threatened by the unrealistic, inappropriate, and outmoded expectations that constitute traditional masculinity. The social indicators point to the massive and deadly cost of sexism to men: men have a 10-1/2 year shorter life expectancy; men have 300% to 500% higher rates of death from murder, accident, alcoholism, suicide, and virtually every major illness; men make up 95% of the state and federal prison population. These and other indicators point to a massive pathology that deserves to be better understood.

I speculate that further research will demonstrate that the traditional concept of masculinity makes work more difficult--more a source of distress--for the individual who has been so socialized. For such individuals, it is more difficult to acknowledge and mediate stress, more difficult to recognize and respond to health and safety hazards in the workplace; more difficult to perform or manage work in a collaborative or participatory manner; more difficult to obtain help from friends or professionals; more difficult to recognize the value of change and respond nondefensively; more difficult to nurture creativity and innovation in self and in others.

We should expect that these effects are likely to be highly significant. Work has played the traditional role in men's lives that men were once supposed to play in the traditional woman's life; it is the external criterion for personal success. Perhaps an increased awareness of and sensitivity to men's issues will help us to improve the health and safety of all who work.

Without additional sensitivity to the effects of sex roles on men, and how men recognize and report their stress reactions, our efforts to understand and effectively respond to occupational stress issues will be partial and inadequate. It is not hard to imagine that job stress would be inaccurately viewed--and tragically dismissed--as a "female problem."

# STEALING TIME: INDIVIDUAL STRATEGIES FOR DEALING WITH STRESS

MARCIA LOVE
Consultant
New York State Workers Project
320 8th Avenue
Brooklyn, New York, 11215

## INTRODUCTION

The high levels of stress endured by office workers have only recently come to light in the scientific community. Researchers have begun to describe and document office stressors and their long-term health effects on workers. Government researchers have also taken an active role in this research, and their findings confirm the concerns expressed by secretarial and clerical workers about their working conditions. In 1977, NIOSH found that secretaries had the second highest incidence of stress-related diseases of 130 occupations studied. And in 1980, NIOSH researchers discovered that VDT operators experienced higher levels of stress than any other occupation studied to date, including air traffic controllers.

Briefly summarized, the major sources of stress found among office workers are these:

1. Job design problems, particularly rapid work pacing, long working hours, repetitive or monotonous work (ref.1); underutilization of skills (ref. 2); lack of decision-making ability in the way tasks are designed or executed (ref. 3).
2. Poor work relationships between employers and employees, for example, lack of recognition of office workers' achievements (ref. 2); too many layers of supervision, conflicting demands of supervisors; electronic surveillance of office workers (ref. 4).
3. Socioeconomic factors, such as low pay, lack of promotional opportunities and raises (ref. 4); the pressures of the dual role of women workers as participants in the paid labor force and as homemakers (ref. 5).
4. Work environment, including poor work station and office furniture design; physical exposures (noise, lighting problems, radiation); air quality/ventilation problems, such as tight building syndrome (ref. 6), drafts, air temperature variation; fire and safety hazards; chemical exposures (ref. 7, 8, 9, 10).

The news that coronary heart disease was twice as high for female clerical and secretarial workers than other female workers points to the serious toll that these stressors are taking on office workers (ref. 2). Other research now underway also suggests an association between office work and cardiovascular problems (ref. 3, 11).

Coronary heart disease is not the only result of the stressors listed above, nor is it the only health effect of stress. A host of other stress-related health problems has been known for decades, including hypertension, ulcers, spastic colon, migraine headaches, and a variety of psychological problems. And it is also known that certain behavioral responses to stress contribute to these and other health problems. The health effects of smoking, alcohol consumption, and drug use -- common behavioral responses to stress -- are widely reported in the literature.

But there are other behavioral responses to stress among office workers. These responses--everyday occurrences--have received little attention in the scientific community, though they have been reported elsewhere. These responses promote health rather than threaten it (as in the case of smoking and drug and alcohol use).

The oral work histories recorded by Studs Terkel (ref. 12), Jean Tepperman (ref. 13), and Barbara Garson (ref. 14, 15), offer eloquent descriptions of the strategies that office workers use to grapple with sources of stress on their jobs. I have continued to research this subject using a variety of means: numerous workshop discussions with public and private sector workers at occupational health conferences; classroom discussions about stress-coping mechanisms with students taking occupational health courses; a review of the popular literature, which includes feature films, documentary films, television shows, and songs about office work, as well as fictional and nonfictional sources; and an open-ended questionnaire.

Though much of the research is anecdotal, based on individual accounts or informal conversations with groups of workers, its potential significance cannot be underestimated. Many occupational health problems have initially been identified in just this way by workers themselves through conversations at lunch, during work, after work, etc. For example, in the 1970's, chemical workers in Lathrop, California, at a plant manufacturing DBCP alerted the nation to the hazards of exposure to this pesticide. Their awareness came through conversations with one another about the low birth rate in their families and led to their subsequent discovery that DBCP made them sterile. Similarly, it was VDT operators themselves who first noted the eye strain, neck strain, back pain, and other symptoms that are now commonly recognized health problems of people doing this type of work. VDT operators brought

---

I developed a questionnaire on mechanisms used to combat stress and circulated it for comment to 200 office workers from around the country. These office workers attended the July, 1981 Summer School for office workers sponsored by 9 to 5. This Summer School is an annual event, held at Bryn Mawr College, Bryn Mawr, Pennsylvania.

their complaints and concerns to the attention of their unions and other organizations, such as working women's groups, and COSH groups. These organizations in turn requested that the government begin to research these complaints. This research ultimately led to the NIOSH findings that VDT operators had the most stressful job of any studied. Employee organizations have also pressed for improvements in VDT machine design, office furniture design, and for changes in office lighting, work station design etc.

Since employees have made important contributions in identifying occupational health problems such as these, it should be of interest to learn what office workers have to say about the ways in which they manage stress at work, as well as the reasons why they employ these techniques. The discussion of stress management techniques will focus on responses to two of the stressors outlined above, job design problems and employee/employer relationships.

## Background

Job design problems vary widely among office workers. While one secretary may be overloaded with work, another may have too little to do. While one secretary may administer the daily operation of a department, another may be given little responsibility. Clerical job content also varies. One clerk might answer telephone inquiries, develop and maintain a record-keeping system, and do other assorted tasks, another may be limited to one task, say key punching, in which activities are limited to a few circumscribed motions.

Employee/employer relations also vary considerably. In a 1:1 relationship between manager and secretary, the interpersonal relations carry heavy weight; they play a defining role (ref. 16). One boss may view his secretary as an "office wife," "servant," "errand girl" and assign tasks to fit the role. Another may view his employee with respect and appreciation. He assigns a reasonable workload, allows the employee autonomy in getting the work done, and acknowledges the employee's abilities. A third may view the secretary as a "work horse," overload her with work, and berate her when the task is not complete. A fourth may underutilize the employee's skills and demean the employee's abilities (ref. 17). The secretary who works for a number of people may also experience the job variously as a source of stress or as a source of pride, depending in part on the interpersonal relations among her and her supervisors. One secretary may be the victim of competing and conflicting priorities and personalities among her supervisors. Another may work in a more harmonious atmosphere.

Supervisory relationships take on a different quality in clerical jobs. For example, telephone operators, key punchers, and other machine operators complain that "Big Brother is watching," because their production quotas and error rates are electronically monitored. Public sector clerks, whose ranks are thinning due to cutbacks in funds for public programs, are pressed to speed up their work. They live under the threat of "being written up" by supervisors who themselves are under pressure from management to produce more work with fewer resources (ref. 18).

## DISCUSSION

### Stealing time

Despite these variations in the ways stress is experienced, office workers have developed a similar set of responses to reduce it. A widespread practice, simply put, is to steal time. This is done in a number of ways. Those who can, get up. They leave their desks, take a walk around the office, go to another floor, to the rest room, to the candy stand, or they find a window (if one is to be found) and look out of it. Other forms of escape include coming in late, leaving early, taking long lunches, and taking a mental health day off from the job.

Motivation for this type of behavior varies. For some, these small acts of freedom mean physical relief from sitting for long stretches; for others, they are a means to avoid the gaze of a supervisor, to find quiet from an overcrowded work station, etc. A loan processing clerk from Seattle who responded to the questionnaire at the 9 to 5 Summer School called these her "minute vacations."

Although secretaries and other office workers such as administrative assistants can legislate their breaks by making a getaway, millions of other workers, particularly machine operators (for example, letter sorters and switchboard operators), cannot. They are apt to have more rigid work rules: they must be prompt, they must punch in, and once at the work station, they must remain there. They are subject to the demands of the machine they operate. A word processor who worked at a San Francisco bank for 5 years commented:

> People felt that they were plugged into machines, that they were appendages to machines rather than people performing functions with other people....We used to have jokes about how we expected that soon they'd chain us to our desks and give us catheters so we'd never have to go to the bathroom.

Her comments are further illustrated by a key punch operator in New York:

> When you punch, your hands are occupied, your eyes are occupied, you can't move your body. Sometimes people used to try to talk and punch at the same time. When I was a supervisor and I'd see it, I'd say, "Just stop and talk for ten minutes. You'll enjoy it more"....I felt you had to let them stop, smoke, go to the bathroom, anything. Otherwise it's just too constant, just gray for hours at a time. No way to break it.

Another common stress reliever besides escape is talking. Once again, as with escapes, the motivations vary: to break the monotony as the key puncher above indicates, to vent frustration, to fill in the long spaces where there's not enough work to do. Again, as in the above examples, there are workers who can't talk. The demands of the job won't permit it.

For those who have access to it, the telephone is, of course, a stress reliever. People call friends and family or conduct personal business. AT&T can attest to the popularity of the telephone as a stress reliever. In 1980, for example, over 285 million (emphasis added) calls were made in New York City alone to Dial-a-Joke, Sportsphone, the horoscope, the weather, etc. AT&T officials indicated (personal communication, July, 1981, by telephone) that they don't have information on when the calls are made, or by whom, but, they said, at least 100 Manhattan-based companies have installed lineblocking equipment, presumably as a cost-cutting measure, so that workers cannot get through to those numbers. AT&T officials then added that workers in some places are retaliating by calling the services in other cities (ref. 20).

Escapes and talking are only two of many time-stealing activities. Other popular time stealers include eating, smoking, and doing personal business. The last item is a catch-all for balancing the checkbook, paying bills, writing letters, making lists of errands, in other words, "You know, doing the things that there isn't time for when you work full time and you got to get home to the kids," as a secretary from Chicago put it.

Other time stealers fall under the category of acts of resistance; for example, work slowdowns, rearranging work assignments to suit the priorities of the workers rather than the boss (ref. 19), even sabotage. Examples of the latter include "accidentally" spilling coffee in the photocopy machine, dropping a paper clip in the typewriter, and misplacing or misfiling files (ref. 21).

Some time stealers are group activities, such as betting pools, raffles, card games, tic-tac-toe games, crossword puzzles, and practical jokes. And, of course, there are parties. Birthday parties, retirement parties, wedding and baby showers, held inside or outside the office on company time, are an integral part of office culture.

### Other behavior

All of the activities described so far offer an escape hatch from work, in one way or other. Some people, however, relieve stress while doing work, not simply by escaping from it. One method to relieve stress is making up systems of rewards and punishments. For example: If I type this page correctly, I'll get up and go to the rest room." Or, "I'll work on filing until 11 o'clock and then switch to something I like better."

Barbara Garson (ref. 14) describes how she devised a system in order to survive a job typing pedigree information at the American Kennel Club. The work consisted solely of typing into columns line after line of dogs -- their names, sex, birth dates, color, parentage, owners, and breeders. She transcribed the information from stacks of IBM cards.

> Minute followed minute like slow drips from a leaky faucet. For the first hundred entries or so I felt that I had done something each time I typed a dog and turned his card. But the feeling grew weaker and weaker. Now, even when I completed a whole pile and reached for a new rubberbandful I didn't feel much....How could I explain to anyone on the outside the little surge of joy when I see owner and breeder's name the same? Or that I've promised myself a break for a tropical fruit Life Saver the next time the day and month of the dog's birthdate are the same?

Another phenomenon is making up games with co-workers, which consciously or unconsciously increase productivity. The New York key punch operator referred to earlier illustrates:

> One thing Aida and I used to do is have races. On the older machines you had to hit harder and they made a louder noise. So we could hear each other, and when we were doing the same job, we could race. Sometimes we'd synchronize. But you're always pressured to go the fastest with the least errors. So we'd synchronize for awhile, but it would always turn into races. We didn't plan the races but we found ourselves listening to how fast the other person was going and doing it a little faster....Aida and I are very good, high performers -- bored, but still high performers. So we thought racing might be our special thing. Then a girl, Janet, told us she did the same thing. I guess it was the only kind of entertainment you could have. Like I said, your hands were occupied, your eyes were occupied, you couldn't move your body, couldn't talk (ref. 14).

Responses to especially stressful situations

In extremely stressfful situations, certain behavior becomes intensified. Questionnaire respondents stated that they ate more, smoked more, and made more phone calls. Another response was voiced by a clerk at Metropolitan Life Insurance Company in New York:

> When I'm having to do several jobs at once, when the agents want a lot of information quickly, especially at bonus time, at the end of the month, I get angry, irritated. Start slamming things around, like the phone. Sometimes I get ornery to my friends at work. And I don't want to.

A co-worker of hers told me: "The pressure is too much. Just too much. You know, sometimes, I go into a closet...and scream!" Other responses to extremely stressful situations include, "going into a room and crying," and writing--and sometimes sending--nasty notes to the boss.

A cashier at the New York State Department of Motor Vehicles described this situation:

> The public is too nasty, too rude. It didn't used to be this way. It's because of the budget cuts. People leave, no replacements. We're so short. We need help....And then the new computer is down. Every other day it seems like....I was saying, the public gets so nasty, I just got to get out. I get someone to cover, short as we are. I walk out that door, take a walk till I cool off.

The same worker related that a co-worker was once so provoked by a customer that she lost control and jumped over the counter and punched him. This was not the only incident of violence reported by public employees who deal with the public under stressful circumstances. For example, assaults or threats of assault on office workers in welfare centers are increasing throughout the country. At least one union has asked the government in its state to enact legislation to ensure the safety of its members.

Much has been written about other responses to stressful situations, such as smoking, drug usage, and alcohol consumption, particularly among workers in blue-collar jobs and among executives. Researchers have also noted similar patterns of behavior in clerical workers. It isn't surprising, then, to learn that the number of women who smoke has increased in the last decade (ref. 22) or to hear of examples of drug and alcohol consumption, particularly among workers performing machine-paced jobs (ref. 22).

<u>Mental responses to stress</u>

Typical answers to the question, "What do you think about to get through the day?" include errands, what to make for dinner, week-end plans, and sex. Some people engage in a thought process that might be termed tuning out. Workers who said that their jobs were not challenging enough or were demeaning or monotonous described what their minds did while their bodies went through the motions. A photocopy machine operator in a New York publishing house, says:

> I press the 'Print' button. The world goes away. For a few minutes I'm gone, I'm out there. Wailing away. Tunes rushing around my head. Sometimes lines from two different songs meet inside, and I try to figure out why they came to me. I analyze the melody. I can't do this when I got to make single copies of something. But it passes the time when I got big jobs to do.

---

The Pennsylvania Social Services Union asked the State Legislature for protective legislation for welfare center employees in November, 1982.

Other workers daydream or free-associate. A telephone receptionist describes how she uses daydreams as an outlet for creative expression:

> I only have a few minutes, if I'm lucky, between calls. So I can't do something like make up stories or write a letter. The phone dictates everything I do. This crummy little machine with the buttons. A machine that a monkey could operate....It's even hard to talk to people at work....I'm always thinking the phone's going to interrupt....So I figure, to pass the little time I have, I think about family, the weekend plans....I like to think about rainbows and mountains. You have to have a good imagination. If you don't and you bore easily, you're in trouble....I write bad poems, sometimes I do drawings. They are reds and blues. Peaceful colors (ref. 12).

While some workers might tune out and others free-associate, others construct fantasies. Get-the-boss fantasies, for example, are not as uncommon or as far-fetched as one might assume from watching the film Nine to Five.

An audit clerk in Cleveland who had a quarrelsome, petty supervisor whom she said was universally disliked, responded to the questionnaire in this way: "I have fantasies of hiring someone to rob her and scare her so she won't come to work." Another clerk wrote that she would like to "move the whole office to another place and not tell the boss....Open up my own place, and run him out of business." A veteran secretary in the insurance industry in Hartford "dreamed of the day when we will be unionized, so the workers will have more weight and get supervisors fired."

And finally, there are workers who handle stress through meditation. A questionnaire respondent illustrates: "I repeat to myself over and over, "It's the job that is making me feel badly." Another summed up her feelings with this exercise: "I'm a worthwhile human being. The boss should recognize my worth."

## CONCLUSIONS

Several conclusions can be drawn from this discussion of survival mechanisms. First, office workers devise a wide variety of stress management techniques during the course of the day. These fall into two large categories: actions and thoughts. Actions encompass a number of time-stealing strategies: physical escapes, conversation with co-workers, using the phone, snacking, doing personal business, etc. They also include acts of resistance (for example, work slowdowns and various forms of sabotage); and organized group activity, such as raffles, games of many kinds, and parties. Other actions occur while doing work, rather than escaping it. These include self-imposed productivity tests, rewards and punishment systems, and the like. Workers also use a number of different types of thought processes to get through the day: for example, free association, fantasies, and meditation.

Second, job design characteristics can greatly influence the type, frequency, and duration of stress management activity of this nature that a worker engages in. In general, secretarial jobs have more task variety and flexibility than machine-paced clerical jobs. The ability to move around the office, talk on the phone, and leave the building to run erands is often a built-in component of a secretary's job, thus affording increased access to stress-relieving activity. Clerical jobs tend to be more confining. Workload also varies more in secretarial jobs than in machine-paced clerical jobs, so the underutilized secretary who is forced to "look busy" has more time on her hands to engage in stress relieving activities than a mail-sorting clerk who must keep a steady pace at a sorting machine.

Third, although secretaries and clerks may not have the same degree of freedom to engage in stress-relieving activity, the desire to get relief from stressful jobs is shared by office workers across the spectrum of jobs.

Fourth, all of these stress relievers seem to be integral features of office work, just as time-outs are essential to sports. Moreover, they are condoned business practices, unless engaged in excessively. For example, workers in many establishments are allowed a few minutes' grace period when arriving at work in the morning, but management may draw the line if a few people are consistently late, or if the few minutes stretches into a half hour.

This last point is worth pausing to consider for a moment. Why are these activities tolerated? On the surface, they are violations of the work ethic, an affront to the goals of efficiency and productivity. They are time wasters. Indeed, any of us would be disciplined or fired if we submitted reports to our superiors saying that we spent time on the phone, taking walks, and paying bills. Yet these practices are tolerated, in part because they reduce a managerial burden. They are a substitute for a planned program to deal with gaps in work flow or problems in workload, size, or other job design problems. And second, because these activities provide us with needed breaks from a long work day or a poorly designed job, they refresh us and, no doubt, very often enhance productivity rather than waste it.

Fifth, most of the stress relievers described here promote health rather than undermine it. For some office workers these activities provide enough relief to get through the day. But in the main, they are not an effective enough force to offset the combined stressors of poor job design, poor employee/employer relations, unhealthy work environments, and socioeconomic difficulties, as we've seen from the data on long-term health effects.

## IMPLICATIONS

Employers across the country, aware of the effects of stress on employee health, have not relied solely on workers' ability to institute their own stress-reduction techniques. Stress discussion workshops, exercise classes, smoking cessation clinics, and other programs now in vogue among executives to curb rates of heart disease, are beginning to filter their way down to clerical and secretarial staff. And, of course, various programs geared toward reducing or eliminating the more problematic responses to stress, such as alcoholism and drug use, have been in place for years. Employee Assistance Programs, for example, are a nationwide phenomenon. Employer-initiated programs such as these have proved somewhat successful in alleviating symptoms to some stress-related health problems. But, like the employee responses described above, they too, are not an effective solution to combating job stress, because they do not attempt to control stressors at their sources. Instead, employers tend to view stress as an immutable fact of life, a permanent condition. Therefore, employees must adapt to stressful conditions by changing their individual behavior, or suffer the physical consequences on their health.

A growing body of literature points to another approach to stress reduction. Some researchers assert that since the causes of stress are external to individuals and social in origin, they require social solutions. These researchers have investigated the value of developing social support mechanisms among co-workers for that purpose (ref. 13, 23, 24, 25, 26). These mechanisms, which take the form of workplace stress discussion groups, union-based health and safety committees, and support organizations outside the workplace (for example, the network of 9 to 5 groups and COSH groups around the country) afford the opportunity for office workers to air with one another their common concerns about the causes of stress and to develop strategies to improve their working conditions. In so doing, these groups have begun to tackle stressors at their sources. Their efforts have resulted in the creation of promotional ladders and changes in job classification systems, substantial upgrading in pay in many union contracts, as well as legal sanctions against sexual harrassment and racial discrimination. They have also been successful in spurring research on work environment hazards and have helped achieve some changes in work station design. These efforts have not been limited to the U.S. Worker-based movements of office workers are pressing for similar improvements in many European countries and have made inroads in other areas, notably increased employee decision-making ability in work organization (ref. 27, 28).

There is still much room for improvement in this country. Employers, government officials, and labor leaders have not paid enough attention to reshaping the office environment to reduce stress, yet the need has never been greater. If anything, current trends in office job design show that stress levels will rise. Indeed, as the 21st century approaches, enormous and rapid changes in office work as a result of computerization are transforming the "office of the future into the factory of the past" (ref. 29). The jobs that will be available to millions and millions of American workers will be highly stressful machine-paced clerical jobs, bringing with them the same problems faced by industrial workers who faced mechanization 150 years ago. Loss of skills, decreased decision-making ability over the work process and product, monotonous and repetitive work, relentless pacing, unmanageable workloads--the results of the application of Taylorist principles in blue-collar work--are now in evidence in the automated office (ref. 4, 15, 19, 27, 28, 30, 31, 32, 33). Secretarial jobs are already beginning to disappear, taking with them the little freedom of movement and decision-making ability that these workers enjoy.

Management, government, and labor must begin serious broad-scale efforts to control the sources of stress at their points of origin--to change the content and structure of office jobs so that they afford some stimulation to human beings who must perform them, and where that is not possible, to reduce workers' exposure to those jobs through such administrative controls as a shortened work day, longer breaks, a shortened work week, task rotation, and the like. The tasks are formidable, but not impossible. We are rapidly approaching a time when it will be technologically feasible to meet the challenges described by the worker below:

> I think most of us are looking for a calling, not a job. Most of us, like the assembly-line worker, have jobs that are too small for our spirit. Jobs are not big enough for people....You throw yourself into things because you feel that important questions--self-discipline, goals, a meaning of your life--are carried out in your work. You invest a job with a lot of values that the society doesn't allow you to put into your work (ref. 1).
>
> Or as Studs Terkel puts it in the introduction to Working: ...the search that many people have for daily meaning as well as bread, for recognition as well as cash, for a sort of life, rather than a Monday through Friday sort of dying (ref. 12).

REFERENCES


1 NIOSH, An investigation of health complaints and job stress in video viewing, Department of Health and Human Services, Cincinnati, 1981.

2 S. Haynes and M. Feinleib, "Women, work and heart disease: prospective findings from the Framingham Heart Study," American Journal of Public Health, 70 (1980).

3 L. Cranor and others, Job characteristics and office work: Findings and health implications, Paper presented at the NIOSH conference on Occupational Health Issues Affecting Clerical/Secretarial Personnel, Cincinnati, Ohio, July 1981.

4 Working Women Education Fund, Warning: health hazards for office workers, Cleveland, April, 1981.

5 Women's Occupational Health Resource Center, Women at work--their dual role, Columbia University, New York, 1980.

6 D. Michaels, Tackling tight building syndrome: What can workers do?, Paper presented at the NIOSH conference on Occupational Health Issues Affecting Clerical/Secretarial Personnel, Cincinnati, Ohio, July, 1981.

7 A. Hricko and M. Brunt, Working for your life: A woman's guide to job health hazards, Labor Occupational Health Program, University of California, Berkeley, 1976.

8 J. Stellman, Women's Health, Women's Work, Pantheon, New York, 1977.

9 M. Love in Women & health issues in women's health care, State University of New York, College at Old Westbury, New York, (1978) 17-22.
10 J. Fleishman, W. Chavkin, M.D. (Ed.) Monthly Review Press, New York, 1983.
11 G. Gordon and B. Snow, Presentation of preliminary findings on stress in open offices, American Psychological Association Convention, 1982.
12 S. Terkel, Working, Avon Books, New York, 1975.
13 J. Tepperman, Not Servants, Not Machines, Beacon Press, Boston, 1976.
14 B. Garson, All the Livelong Day, Doubleday & Company, New York, 1975.
15 B. Garson, Scanning the Office of the Future, Mother Jones, 6:6 (1981) 32-41.
16 R. M. Kanter, Men and Women of the Corporation, Basic Books, New York, 1977.
17 M. K. Benet, The Secretarial Ghetto, McGraw-Hill Book Company, New York, 1973.
18 M. Love and C. Wintle, Out of title work in the administrative services unit, prepared for the New York State Committee on Work Environment and Productivity, Albany, New York, September, 1982.
19 H. Downing, The making of twentieth century servants, in Working for Capital: Case studies in Working Class Habituation and Resistance, Rontledge & Kegan Paul, London, 1983.
20 T. Harper, Dial-It Services Mean Big Revenues, Middlesex News, Middlesex, New Jersey, May 25, 1981, pp. 6-7.
21 M. Belkin, Drowning in the Steno Pool in Liberation Now, Dell Publishing Company, New York, 1971, pp. 77-82.
22 R. Howard, Strung out at the phone company: How AT&T workers are bugged, drugged, and become unplugged, Mother Jones, 6:7 (1981) 39-45, 54-59.
23 L. Schore, Stress Groups: A union based approach, paper presented at the NIOSH Conference on Occupational Health Problems Affecting Clerical/Secretarial Personnel, Cincinnati, Ohio, July, 1981.
24 Working Women, Office Work in America, Cleveland, April, 1982.
25 University of Connecticut, New Directions Program, Medical Center, unpublished, Farmington, Connecticut, 1981.
26 R. Karasek and others, "Coworker and supervisor support as moderators of associations between task characteristics and mental strain, Journal of Occupational Behavior, (1981) 181-200.
27 J. Gregory, Technological change in the office workplace and implications for organizing, in Labor and Technology: Union Response to Changing Environments, Donald Kennedy and others (Eds.), Dept. of Labor Studies, the Pennsylvania State University, Pennsylvania, 1982, pp. 93-102.
28 O. Ostberg, The Empirics of specialization and division of labour among Swedish salaried employees--A trade union view on the technological development, paper presented at International Symposium on Division of Labour, Specialization and Technical Development Linkoping, Sweden, June 7-11, 1982.
29 K. Nusbaum, in Office hazards: How Your Job Can Make You Sick, Tilden Press, Washington, D. C. 1981.
30 H. Braverman, Labor and Monopoly Capital, Monthly Review Press, New York, 1974.
31 E. N. Glenn and R. L. Feldberg, Proletarianizing clerical work: technology and organizational control in the office, in Andrew Zimbalist (Ed.), Case Studies in the Labor Process, Monthly Review Press, New York, 1979, pp. 51-72.
32 J. Greenbaum, In the Name of Efficiency: Management Theory and Shopfloor Practice in Data Processing Work, Temple University Press, Philadelphia, 1979.
33 M. Murphree, Rationalization and satisfaction in clerical work: A case study of Wall Street legal secretaries, Paper presented at the Technology and Labor Conference sponsored by Monthly Review Foundation, New York City, April 3, 1982.

# OCCUPATIONAL STRESS: A UNION-BASED APPROACH

LEE SCHORE
Institute for Labor and Mental Health
3137 Telegraph Avenue
Oakland, California 94619

## INTRODUCTION

The last 5 years have seen an increase in research and a growing body of knowledge developed to help us deal with the problems caused by stress, but stress at the workplace remains a sensitive issue. Though popularized through the media, it is still not legitimized, still not accepted as a "real" hazard. No guidelines are available to measure it, compensation for it is rare, and little systematic work has been done to develop strategies for preventing stress at the workplace.

In the past few years, we have attempted to address some of these issues, an effort which in itself requires considerably more research. This paper presents what we believe to be an exciting and important direction in dealing with the occupational stress of clerical workers as well as the rest of the work force.

## THE UNION-BASED APPROACH TO STRESS REDUCTION

The Institute for Labor and Mental Health has been working with trade unions in northern California since 1977, developing programs to address the fundamental issues of workplace stress. We see union involvement as critical in combating stress and a union-based approach as essential to creating healthy workplaces. We came to the labor movement as psychotherapists and union activists who shared a concern with local unions about the physical and emotional well-being of their members. Together, we have worked on many projects: presentations to health and safety committees documenting the presence of stressors at the workplace, consulting on stress-related grievances and disabilities, as well as organizing educational workshops, clinical counseling services, occupational stress groups, and steward training sessions. We are now in the final year of a 3-year grant sponsored by the National Institute of Mental Health (NIMH) to train shop stewards from the Service Employees International Union (SEIU) and the Communication Workers of America (CWA) in skills to co-facilitate our occupational stress groups.

The dominant approach to stress reduction developed from a focus on executive and managerial stress. This approach, however, does not adequately take into account the conditions of clerical, service, and blue-collar workers, nor has it developed remedies that are as useful to these workers.

It is an approach that remains focused on treating the individual and reducing the effects of stress that are felt by that individual. While it utilizes some valid and important techniques that can reduce the physiological symptoms of some stress reactions, it generally does not look beyond the individual to identify the sources of stress. It emphasizes a good person-environment fit, but the person is expected to do the fitting. Sophisticated tests are being developed so that corporations can presort what kinds of personality types will fit into existing job categories. There is no comparable attempt to change the jobs to fit the needs of the existing human beings. While we believe that individuals can be helped to find relief from the effects of stress, any attempt to reduce stress must act to reduce or mitigate the stressors in the environment.

The union approach is based on the belief that job stress is related to the structural conditions of the workplace and affects all the workers there, not merely the troubled worker. We do not seek to help workers manage their stress better or to adjust to the conditions that are potentially dangerous to their health. Rather, when we speak about stress reduction we are really talking about stress prevention--creating healthy workplaces. This requires collective solutions and collective actions to reach those solutions.

The occupational stress group is the format we have developed specifically to deal with the stress of clerical, service, and blue-collar workers. The choice of the group format is neither random nor merely a matter of convenience but follows directly from our understanding that occupational stress is a social problem, not an individual problem.

Work is indeed the defining aspect of one's life. Our sense of ourselves is very much determined by the work we do and the effect that work has on us. The kinds of jobs we have and the status and rewards that society has placed on those jobs are some of the primary ways we understand how others judge us, as well as the way that we judge ourselves and measure our self worth.

## THE SOCIAL ASPECTS OF STRESS

Any phenomenological account of the impact of the workplace will have as its core the issue of human dignity. When we look at the factors that create stress reactions, we find that many of the stressors relate directly to workers' lack of opportunities to use their intelligence, creativity, and skills, as well as the degree to which they feel they are not given respect and dignity as human beings. Thus, workers must not only deal with the stress caused by an oppressive or unrewarding workplace, with its threats of automation, machine pacing, noise, eyestrain, and so on, but also deal with the stress of a negative self-image caused by a work environment that underutilizes or even denies their capacities. This latter form of stress often makes workers believe that the feelings of strain they are experiencing is in some way their own fault.

This dynamic of self-blame is a very powerful one, and it is within this context that stress becomes even more destructive. In many ways self-blame is the loom upon which the fabric of stress is woven. Workers who face daily the reality of jobs that repress their every impulse to use their own knowledge and judgment frequently internalize their anger and blame themselves for the situations they find themselves in.

Key to the sense of self-blaming is the incredible power still exercised by the widespread belief that this is a meritocracy: that in America anyone can still make it if he or she really tries and that the individual has control of his or her destiny. The converse to this, then, is that if an individual is in an oppressive work situation it is because he or she was not good enough to get a better job. Self-blaming is reinforced by the attitude of management that often sees the problems of the workplace as mainly the fault of the workers, whom they may consider lazy, careless, unmotivated, disloyal, and greedy. Management's solution, all too often, lies in workers changing their attitudes.

The picture is further complicated by the very nature of stress itself, which is manifested in so many different forms that it can easily be accepted as a reflection of some individual weakness or personal flaw, rather than different individual responses to a common condition. Any five workers in the same office exposed to and reacting to the same stressors might have five different symptoms: one may have headaches, one colitis, another allergies, yet another low back pain, and the fifth insomnia. All five may also experience the consequence of this stress in nonwork situations: one may be going through a divorce, another's children may be in trouble at school, yet another may feel socially isolated with few friends, one may be slipping into habitual drug dependence, and the fifth may be suffering from anxiety and depression.

Though these five workers may be aware of each other's complaints and some of their problems, and they are all aware of their own dissatisfaction with their job, they do not necessarily see them as connected, but rather continue to feel them as individual and isolated conditions. The tendency is often to blame oneself as not being able to cope.

The impact of this kind of thinking is the creation of what Lerner (ref. 1) calls "surplus powerlessness," a sense of powerlessness greater than the objective lack of power that does exist. This is an internalized sense of futility and frustration that becomes a causal factor in the reason why workers fail to engage in actions that are objectively in their self interest. Gardell (ref. 2) identifies this as social helplessness: a passivity that curtails the individual's ability to develop active relations during his or her spare time and to take part in organized goal-oriented activities outside of work. This is reflected in the general apathy that many workers experience in society, and may contribute to the low level of involvement by union members in their unions.

In support of this theory, our contacts with union members have revealed that one of the main reasons given for minimal participation in union activity was that they felt too heavily stressed by work. In many ways this is a circular problem that calls for some mechanism to buffer the impact of stress on workers and potentially free them to use their energies to deal with more structural changes in the conditions that cause stress at the workplace.

The research (ref. 2, 3, 4) points out the importance of social support systems and emotional networks as buffers of stress. Occupational stress groups are a form of social support that can offset some of the psychodynamics of stress and strengthen the ability of participants to develop their own natural support systems at work, in their families, and in the community.

## THE OCCUPATIONAL STRESS GROUP

Insofar as self-blaming and anger are the common resultants of stress, stress groups are effective in reversing the notion that stress is an individual problem, enabling workers to see that the stresses they face at work are not theirs alone, but are shared by others and are rooted in circumstances that are not their fault. Through this understanding, individuals gain an increased sense of self-worth and begin to overcome the process of selfblaming and powerlessness described above.

The occupational stress group is neither a therapy group nor a sensitivity training group. It is neither a problem-solving group nor a class. It is an opportunity for workers to come together, to get information, and to learn to use that information to understand the impact of work stress on their personal lives. The central dynamic of the group is the interaction among the participants as they share the emotional realities of their work experiences and recognize that what they considered personal problems are common shared reactions that are not unique to them, to their jobs, or even to their particular workplaces.

The groups meet once a week for 8 weeks, and the sessions are generally 2 hours in length, though that can vary from 1 hour for lunch-time groups to 3 hours in the evening. Every session teaches a relaxation technique such as breathing exercises, autogenic relaxation, and guided visualization. They also include information on nutrition and exercise. We feel these are useful tools to learn but are careful not to create the illusion that they will get rid of the stress.

Each session includes a short presentation by the facilitator about some topic covering an aspect of stress at work and its impact on personal life. The presentation leads to a detailed discussion by participants about their own situations and the ways that the analysis presented either does or does not reflect their own experiences.

The presentations are not lectures, and there is no expectation that a specific content must be covered in one session. The topics are meant to give a focus to the discussion, but the discussions are generated by the participants themselves as they explore their own lives in relation to the topic.

TOPIC OUTLINE FOR THE 8-WEEK STRESS GROUP

Week One: Introduction to Stress and to Each Other
Week Two: The Physiology of Stress
Week Three: The Psychology of Stress
Week Four: The Organization of Work
Week Five: Self-Blaming and Anger
Week Six: Stress and Discrimination on the Job
Week Seven: Bringing Stress Home
Week Eight: Changing Stressful Working Conditions

We have offered groups for almost 2 years to a broad range of workers from all sectors of the work force. There have been at least 30 groups with 5 to 10 participants in each. Some groups have been limited to a particular workplace or union, and a few groups have been designed for specific populations such as women's groups or retiree groups, but the majority of the groups have had representatives from a variety of jobs and workplaces. The age of the participants ranges from the early thirties up to and through retirement age with very few younger workers. Most groups are almost evenly distributed by sex, and almost every group has been multiracial.

People have come to the groups for different reasons. Some have already been damaged by the effects of stress, others are afraid they will be affected, and others are aware of stress as an issue and want to learn more about it for their unions. And just as they came for different reasons, the immediate effect has differed from person to person. For some, the importance of the group has been simply in having a place to vent their frustrations and anger and have it accepted; for others it has been in having their own feelings validated ("knowing that they weren't crazy after all"), for others it was no longer feeling alone; while for still others it was finally having an explanation that "made sense in their lives." And for some it was understanding how to begin to develop strategies to affect their workplace. Almost all the participants indicated that they felt better about themselves, stronger, and more able to relax.

Typically, as individuals in the group felt an increased sense of selfworth and solidarity with others, they became more actively involved in their unions or in some other area of collective action including their churches or their own family units. Many participants reported back to the group conversations they had had with fellow workers about the things they had learned in the group and were surprised by the positive response they received. A few felt the conversations with co-workers had actually led to better and more cooperative working relationships among them. In every group, there were one or two who began going to union meetings again, a few who became stewards, and several who became involved in health and safety committees or helped their union to sponsor workshops on stress for their members.

These groups were developed as only one part of a training grant and have not yet been the subject of systematic research. However, the response of the approximately 200 workers who have thus far participated in these groups seems to justify further study. The research on social support has called for a practical format to test the theoretical work; we are excited by the promise of the stress group as one such format.

This is a time when we must work together as union members, researchers, practitioners, and management to ensure that the progress that has been made is not lost to Federal cutbacks. Everyone will benefit from work conditions that are rewarding, safe, and productive. Industry will profit from less absenteeism, fewer accidents, and reduced alcoholism and drug abuse. Unions will be strengthened and revitalized. And, individual workers will regain a sense of dignity and self-esteem. The demand for a healthy work environment is one of the central issues of our time.

REFERENCES

1 M. P. Lerner, Surplus Powerlessness, Social Policy, January-February 1979.
2 B. Gardell, Psychosocial aspects of the working environment, The Swedish Information Service: Working Life In Sweden, No. 1, October, 1977.
3 J. S. House, Work Stress and Social Support, Addison-Wesley, Massachusetts, 1981.
4 M. Frankenhaeuser, Do women cope with stress better than men? The Swedish Information Service: Social Change in Sweden, No. 20, October, 1980.

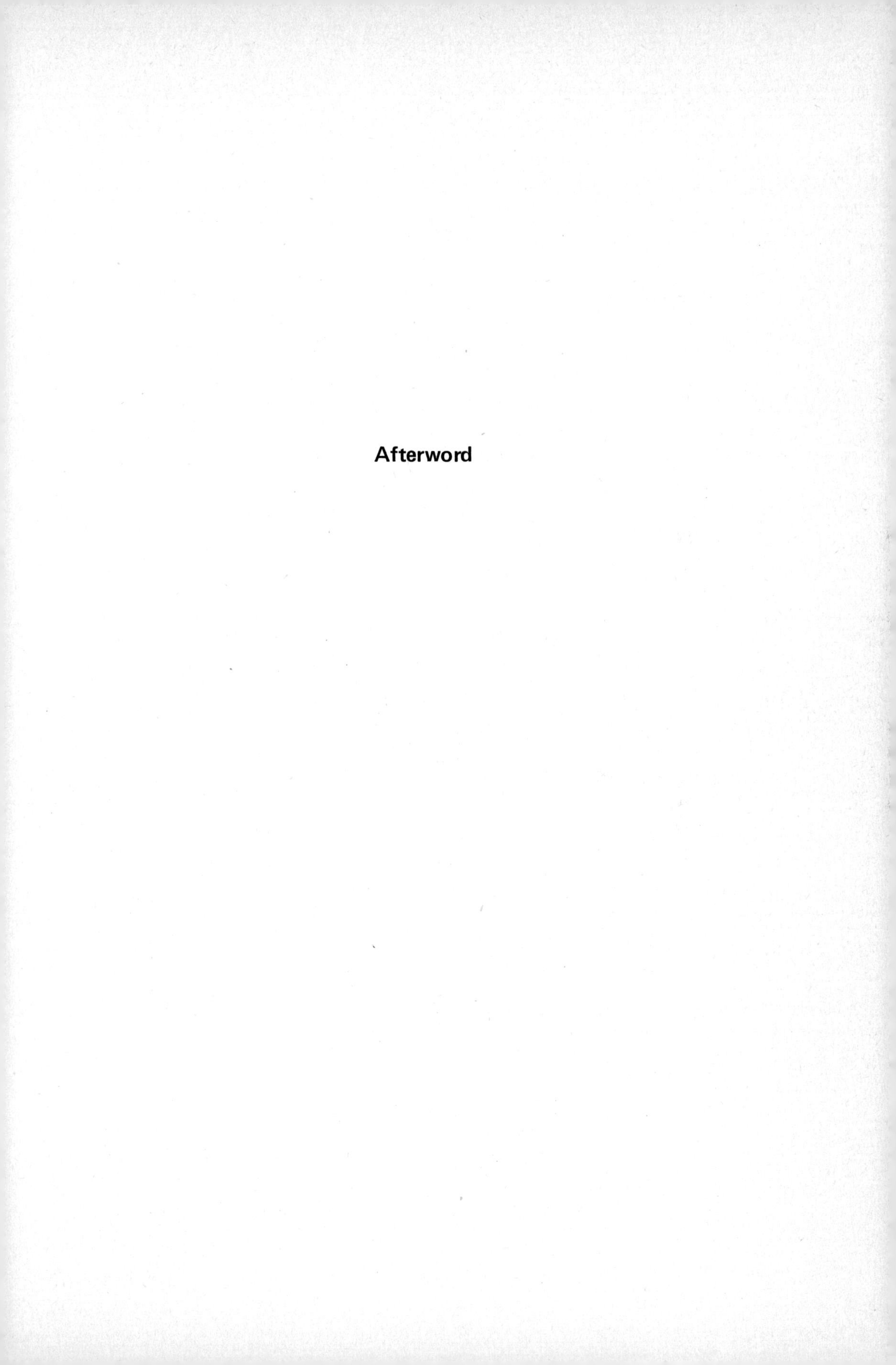

# Afterword

# AN OLD PROBLEM IN A NEW ENVIRONMENT: SAFETY AND STRESS IN THE MODERN OFFICE

GLENN E. WATTS
Communications Workers of America
1925 K Street, N. W.
Washington, D. C. 20006

The safety and health of American workers have historically been a major issue and concern of the labor movement. In fact, health and safety concerns in the workplace were driving forces behind the birth of the labor movement a hundred years ago.

Today, as organized labor enters its second century, that commitment to the safety of workers has never been stronger. As working environments are radically transformed by ever-changing technologies, labor's commitment to the health and safety of American workers takes on even greater consequence.

New technology and new work environments carry with them new and alarming health and stress problems for American wage-earners. We may be on the verge of the paperless office, but we have yet to devise the pressureless office.

The Communications Workers of America (CWA) represents hundreds of thousands of workers in the telecommunications industry. The workers in this industry are probably as close as anyone to the technological revolution sweeping the workplace.

Yet, at the same time, here is an industry that provides proof that modern technology is no cure-all for the problems associated with job stress and job safety. Upon examination of the telecommunications industry, we see that this new wave of technology brings its own special brand of tension and job pressures to the workplace.

Consider the telephone switchboard rooms of yesteryear--long, drab rooms replete with rows of telephone switchboards where as many as 100 operators would work. There was no air conditioning, no curtains, and much of the bulky telephone equipment of that day was totally exposed. For every eight or ten operators, there would be a supervisor peeking over shoulders in the name of greater worker efficiency. If an operator was not processing what was arbitrarily deemed to be an appropriate number of calls in a certain period, a supervisor was quick to point out the employee's shortcoming. And therein lay a major source of job pressure for the telephone operators of years past. We called it oversupervision then. But make no mistake, this was job pressure of the most severe sense. Contrast that working environment with today's ultramodern operator stations and, at first glance, you will certainly marvel at the differences, at the technological advances that have taken place. The deplorable working environment of the telephone operator of yesteryear has been replaced with the bright openess of modern office decor.

The rooms are air conditioned, the floors are carpeted, the equipment is compact and concealed. To the innocent bystander, the workplace of today's telephone operators could not appear more serene or safe.

There is no question that the working environment of the telephone operator has changed drastically during the past 30 or 40 years. What has not changed is job pressure. Modern technology has given us the ability to talk with people halfway around the world in a matter of seconds, but it has not provided us with an answer to job-related stress.

The main source of job stress for telephone operators used to be the constant badgering applied by numerous supervisors. Today, the badgering of the switchboard supervisor has been replaced by the buzzing of the computer. Supervisors measured an operator's performance by the number of calls the operator handled during a given time. The tension resulting from such oversupervision was often overpowering.

The performance of telephone operators is still basically measured by counting the number of calls processed during a given time. Today, however, that performance--or average work time, as it is called--is measured by a computer. But the pressure to perform lives on and job stress unfortunately remains a part of the job.

Research is just beginning to point out that the levels of stress associated with modern technology and the modern workplace are serious, and demand urgent attention from both labor and management. As a result of new technology, computerized measuring sticks for job performance are being rapidly introduced into the workplace. The end result is a work situation where people are under even more pressure.

Nowhere is this new concern over worker safety and stress more appropriate than in today's modern office. Yet, many people have unknowingly categorized office work as being safe, clean, and generally free of stress and pressure. Hence, the term white-collar workers. But just as technology has changed the face of the typical office, studies are now beginning to show that office work can, and many times does, involve serious health hazards and alarming levels of job-related stress.

The perils associated with office work have not been fully documented--far from it. In fact, this is an area where research is still in its initial stages. But the findings to date deliver similar and alarming messages: Office work frequently produces high levels of worker stress and the office environment is far from the haven of safety that many people envision.

In recent years, we have witnessed a dramatic change in the American work force. Clerical workers have replaced manufacturing workers as the largest single segment within the work force. More than half of all the new jobs created during this decade will be white-collar jobs. Some experts have even suggested that 50 years from now, as much as 80 to 90% of the work force could be comprised of jobs involving information processing--white-collar jobs.

As this transition from the factory to the office continues, concern over the working conditions that will confront the ever-growing number of white-collar workers is coming into focus. Initial research has already dismissed one commonly held myth about the office environment. According to that myth, only top-level executives and people who make weighty decisions experience the effects of job-related stress. However, it is not just the highly-paid executive who suffers from heart disease as a result of job pressures. The executive's loyal secretary who lives in a working environment packed with similar pressures is also a prime candidate for a heart attack.

The old idea that "stress is all in your head--it's nobody's fault but your own if you let things get to you" is as outdated as the manual typewriter. Medical researchers have implicated job-related stress as a major risk factor in heart disease. And recent research has further documented that the potential for heart-related maladies is an ever-growing danger in the office.

The Framingham Heart Study (ref. 1), found that women clerical and secretarial workers developed coronary heart disease at nearly twice the rate of other women workers. The potential magnitude of such a problem hits home when one considers that more than 20 million women presently hold office positions in the United States. The American Academy of Family Physicians (ref. 2) found, in a study of six different occupations, that two-thirds of the clerical workers polled described their work as usually or always stressful.

A NIOSH study (ref. 3) of a San Francisco office found higher levels of job stress among clerical operators of video display terminals (VDTs) than in any occupational group ever studied by this reputable organization, including air traffic controllers.

A recent report on job stress in the office included an account of a young woman office worker who was stricken with heart disease at the age of 27. The woman's doctor has concluded that her heart condition was brought about by the pressure-packed environment in which she worked.

With a college background in accounting, the young worker was promised by the insurance company that the switchboard position would be only temporary. Shortly after her work began, the company asked the new employee to help out in the mail room in addition to her switchboard duties. Before long, both jobs became permanent. And so did the job pressure. Several months later, the woman's doctor diagnosed the source of her continuing health problems as a heart condition--a condition the doctor believed was a direct result of continual pressures in the woman's working environment.

The young worker described her plight in this manner:

> Two years after having so confidently begun my career, I am still working at the switchboard. I am stuck in a dead-end job. My medical benefits are exhausted. I make under $10,000 a year. And I have permanently lost my good health.

This account illustrates the seriousness of job-related stress in the modern office. It is a problem that can no longer be tossed aside or taken for granted. New ideas and attitudes will be required as modern technology brings drastic change to the working environments we have become accustomed to over the past several decades. Someone recently commented that the technological revolution would make the Industrial Revolution look like a Sunday School picnic. As change sweeps the workplace, both labor and management must actively seek ways to combat the problems associated with on-the-job pressures. A failure to do so will seriously hamper the ability of workers to adjust and prosper in the workplace of the future.

We in CWA recently joined hands with several managements in what we believe to be one type of revolutionary approach necessary to confront radically changing working environments. This new concept is often referred to as quality-of-work life. These programs are based on greater worker involvement in the decision-making process of the workplace. The theory is that workers will do a good job, enjoy their work, and make the right decisions if they are given the proper chance. In the process, we believe programs such as our quality-of-work life campaign can be invaluable in eliminating many sources of job stress. Greater worker participation in the decision-making process can only improve workers' attitudes about their jobs and increase their levels of job satisfaction.

If we can enhance job satisfaction in the years of change ahead, we will have taken a giant step in preventing many new forms of job stress that accompany modern technology.

While job stress is unquestionably a source of increasing concern to today's clerical worker, other aspects of the modern office pose similar threats to the health and safety of workers. Less-than-ideal lighting arrangements, constant sitting in poorly designed chairs, and office air polluted by irritating fumes from modern office machines are now beginning to be recognized as potential health hazards. When office workers are surveyed about office environments, one of the strongest and most frequent complaints is always poor air quality and poor air circulation. For example, a recent survey found that 90% of the office workers polled complained that their offices were either too hot or too cold.

Because of today's commitment to energy conservation, many buildings are being constructed or redesigned in a manner that prevents adequate ventilation. This sealed environment means that harmful chemicals, dust, cigarette smoke, and bacteria are recycled over and over through the building's cooling and heating system.

Another problem of the modern office--and one of the most serious--involves the potential health hazards produced by video display terminals (VDTs). The health problems associated with VDTs provide the clearest example of how modern technology brings its own special stress and strain to the workplace. In general, VDT operators complain about visual and postural problems--irritation and soreness of the eyes, and discomforts related to the neck and back. After spending several hours in front of a VDT, workers are also likely to experience dull and lingering headaches.

Health problems associated with VDTs are of particular concern to CWA since hundreds of thousands of our members will be using VDTs extensively in the years ahead as more employers look to modern technology to improve their operations. With that in mind, CWA is striving to develop programs that will alleviate the stress and strain-producing factors associated with video display terminals.

It is imperative that manufacturers and employers pay closer attention to the health and safety of workers. The work environment should be designed to fit the needs of workers and not vice versa.

Employers must be constantly reminded that efficient production is best attained when workers are provided with optimal working conditions. Office environments should be properly illuminated with ceiling lights arranged to prevent glare and visual discomforts, including VDT screen reflections. Video display terminal desks should be positioned so that lighting can be controlled by the individual operator. Specially designed VDT desks and chairs should be used to minimize problems related to posture. When VDTs are introduced into the office, employers should arrange for eye examinations to correct the vision problems of the potential operators. Researchers have suggested that such examinations should be made available at least once every 5 years for employees 45 years of age and older. Whenever necessary, the employer should provide special work glasses for all VDT operators. Furthermore, employers should consider job rotation and frequent rest periods as a means of reducing job stress and strain related to the operation of VDTs and other components of the modern office.

CWA's concern with regard to the stress and strain in the modern office represents the newest phase of our union's total commitment to the health and safety of workers in all working environments.

In 1978, CWA established its own Occupational Safety and Health Program and hired a professional staff person to administer the program on a full-time basis. The program has been immensely successful--we have helped CWA locals confront and correct some 250 safety and health problems in the workplace.

CWA's Occupational Safety and Health Program and similar efforts by other unions have had a strong and positive impact on safety in the workplace. Unfortunately, today we find these important programs threatened by new government policies generated in Washington, D.C. It has been more than 10 years since the landmark Occupational Safety and Health Act became law to "assure every working man and woman in the nation safe and healthful working conditions." If the first 2 years of the Reagan administration represent an accurate gauge, it would appear that the health and safety needs of workers will be considered only as an afterthought by this administration.

It is truly appalling to see efforts to end brown lung disease questioned and plans to control lead poisoning being delayed, all in the name of regulatory reform. Yet, this is the administration's approach toward occupational safety and health programs. Apparently, this viewpoint takes its cue from the interests of big business--the interests that have constantly fought OSHA programs over the past 10 years, contending they were too strict and costly to employers. But what price tag can we put on human lives?

How can anyone contend that worker safety and health programs are too costly when some 14,000 workers are killed in accidents on the job each year, and another 100,000 workers die annually from various occupational diseases?

There can be no price tag on worker safety programs. That was the opinion of the U. S. Supreme Court when it ruled that the insensitive cost-benefit analyses of the Reagan administration could not be the basis for determining the effectiveness of occupational safety and health programs.

Washington columnist Richard Cohen did a masterful job of analyzing the administration's approach to OSHA programs following the Supreme Court ruling. Mr. Cohen appropriately wrote:

> "...the Reagan administration had...argued that sometimes it just costs too much to protect the health of American workers.
>
> "In other words, the contest is between dollars and human health--an unequal contest at best. Without government intervening, dollars would always win.
>
> "What is appalling here is not the industry's mentality, but the government's endorsement of it. Industry is industry and left to its own devices and conscience, it will try to increase its profits any way it can.
>
> "To the worker, his life and his health are priceless. There can be no balance. There can be no other considerations.

"After all, a businessman who loses his business can start all over again. A worker who loses his health has lost pretty much everything."

When it comes to the safety and health of workers, the government must take a stand. There can be no choice between benefits and costs when a human life is at stake. Our commitment to safe working environments is more important today than ever before as workers are confronted with rapid technological changes.

The modern working environment brings with it new and dangerous problems with the power to affect the safety and health of workers. As we have seen, modern technology has not found a cure for job stress and job strain. The influx of new technologies and new working conditions means that our work on behalf of worker safety is only beginning.

This is no time for government, or labor, or management to shirk its responsibility to the safety and health of America's workers. Our efforts must reflect a new level of intensity and dedication to cope with the changing workplace. The lives of American workers hang in the balance. There can be no price tag and no limit to our efforts.

REFERENCES

1 S. G. Haynes, M. Feinleib, Women, work, and coronary heart disease: Prospective findings from the Framingham Heart Study. American Journal of Epidemiology, 111, 1980, 37-58.
2 American Academy of Family Physicians, Life style in different occupational groups, Report of a survey. Kansas City, Missouri, 1979.
3 M. J. Smith, B. G. F. Cohen, L. W. Stammerjohn, A. Happ, An investigation of health complaints and job stress in video display operations. Human Factors, 23, 1981, pp. 387-400.

SUBJECT INDEX

AUTHOR INDEX

Rapoport, R., 189(11)193